HIGH-PERFORMANCE AUTOMOTIVE FUELS & FLUIDS

Jeff Hartman

First published in 1996 by Motorbooks International Publishers & Wholesalers, 729 Prospect Avenue, PO Box 1, Osceola, WI 54020-0001 USA

© Jeff Hartman, 1996

All rights reserved. With the exception of quoting brief passages for the purposes of review no part of this publication may be reproduced without prior written permission from the Publisher

Motorbooks International is a certified trademark, registered with the United States Patent Office

The information in this book is true and complete to the best of our knowledge. All recommendations are made without any guarantee on the part of the author or Publisher, who also disclaim any liability incurred in connection with the use of this data or specific details

We recognize that some words, model names and designations, for example, mentioned herein are the property of the trademark holder. We use them for identification purposes only. This is not an official publication

Motorbooks International books are also available at discounts in bulk quantity for industrial or sales-promotional use. For details write to Special Sales Manager at the Publisher's address

Library of Congress Cataloging-in-Publication Data
Hartman, Jeff.
 High-performance automotive fuels and fluids/Jeff Hartman.
  p. cm. —(Motorbooks International powerpro series)
 ISBN 0-7603-0054-2
 1. Motor fuels. I. Title. II. Series.
TP343.H345   1996
629.25'38—dc20   96-8845

On the front cover: Indy Car racer Bryan Herta refuels with pure methanol during the 1995 Texaco 200 at Elkhart Lake, Wisconsin. Methanol is well established as a racing fuel, particularly in Indy Car and drag racing. Methanol produces greater power than gasoline, because a given amount of air drawn into an engine can burn much more methanol than gasoline. *Richard Dole*

Printed and bound in the United States of America

Contents

Chapter 1	**Fuels**	4
Chapter 2	**Chemistry Lesson**	9
Chapter 3	**Lubricants**	15
Chapter 4	**Engines and Fuels**	25
Chapter 5	**Gasoline**	33
Chapter 6	**Aviation Gasolines**	45
Chapter 7	**Gasoline as a Racing Fuel**	55
Chapter 8	**Alcohols**	65
Chapter 9	**Nitro, Monopropellants, and Rocket Fuels**	73
Chapter 10	**Natural Gas and Propane**	77
Chapter 11	**Diesel**	87
Chapter 12	**Antiknock Additives and Water Injection**	93
Chapter 13	**Nitrous Oxide Injection**	99
Chapter 14	**Dual-Fuel Injection**	105
Chapter 15	**Gear Oils, ATFs, and Other High-Performance Fluids**	109
	Glossary	114
Appendix A	**Summary of Data on the Knock Ratings of Hydrocarbons**	122
Appendix B	**Racing Fuels Specifications**	125
	Index	128

Fuels

This book is designed to consider the advantages and disadvantages of various types of internal combustion engine (ICE) fuels and lubricants from the point of view of maximizing power in reciprocating engines, consistent with playing by the rules. The rules include not only sanctioning bodies' specific fuel requirements but, for street vehicles, safety and emissions requirements. As we will see, there is no ideal fuel. That is, there is no single fuel that embodies all the best characteristics of specific energy, resistance to knock, handleability, safety, economics, etc. Hence, there is ongoing research in the quest for better fuels.

For the purposes of this book, there are eleven main fuels: iso-octane, methanol, ethanol, nitromethane, nitropropane, gasoline, diesel, propane, natural gas, M85, and hydrazine. Iso-octane is sometimes a component of gasoline, which is usually a complex blend of various hydrocarbons. But it is useful to consider iso-octane and certain other compounds that are sometimes blended into gasoline as distinct fuels. M85, a blend of 15 percent street gasoline and methanol, is also worth considering as a distinct fuel.

Almost any combustible compound that is a liquid or gas has been tried to power the reciprocating spark-ignition engines. The list includes everything from the fumes from smoldering charcoal to rocket fuels like hydrazine and includes less common liquid fuels like acetone, benzene, nitrobenzene, nitropropane, and ether. Again, no fuel is the ideal fuel, but all of the above eleven main fuels excel in some high-performance characteristics.

Alcohols and nitromethane are common in the higher classes of drag racing, as well as in Indy car racing, various drag boats, hydroplanes, and cigarette boats. The high specific chemical energy of an oxygenated fuel like nitromethane allows it to produce far more power than conventional hydrocarbon gasoline fuels for short duration drags where severe engine stress need only be managed for a few seconds. Good engine cooling properties, high octane, and good specific energy make alcohol fuels the accepted fuels for sustained performance Indy-type racing. Nitro and methanol are often blended for circle track racing. Gasolines are used in almost every type of racing and are sometimes fortified with oxygenated and nitro-bearing fuels for more power. Water or alcohol injection have been used to improve the antiknock characteristics of gasoline and other fuels and to improve engines' volumetric efficiency. Oxidants like nitrous oxide or even pure oxygen have been used to make better use of gasoline's extremely high heating value.

Fuel Characteristics—Definitions

1). Oxygen Requirements. How much fuel can be burned by the oxygen drawn into the motor? Put another way, consider the chemically correct air-fuel ratio (also referred to as the stoichiometric ratio) for the fuel. The lower the stoichiometric air-fuel ratio, the more fuel can be delivered to the

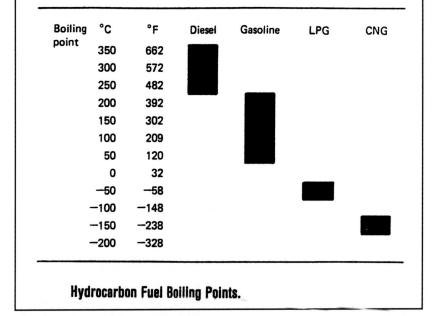

Hydrocarbon Fuel Boiling Points. In general, hydrocarbons with more carbon atoms (higher atomic weight) have higher boiling points. *Reston Publishing Co.*

High-Performance Fuels

Property	Iso-octane	Methanol	Ethanol	Nitrometh.	Gasoline	Diesel
Formula	C_8H_{18}	CH_3OH	C_2H_5OH	CH_3NO_2	N/A	$C_{12}H_{26}+$
Mole. Wt.	114	32	46	61	N/A	170
Oxygen Content	0	49.9	34.8	52.5	0	0
Stoich. air-fuel ratio	15.1:1	6.45:1	9.0:1	1.7:1	14.6:1	-
Heating Value (BTU/lb)	19,100	8,600	11,500	42.7 (MJ/kg)	19,000	18,000
SE at stoich AFR	2.9	3.08	3.00	2.3	2.92	?
Heat of Vapor	0.27	1.17	0.93	0.56	0.18	?
Boiling Point (C.)	?	65	78	?	35-210	190
RON	100	109	109	?	90-100	-
MON	100?	89	90	?	80-90	-
Flame Speed (cm/sec)	34.6	?	?	?	34-40	?

Property	Propane	NG (Methane)	M85	Hydrazine	Nitroprop.
Formula	C_3H_8	CH_4	85% methanol	N_2H_4	$C_3H_7NO_2$
Mole. Wt.	44	16	N/A	32	89
Oxygen Content	0	0	30.6	0	35.9
Stoich. air-fuel ratio	15.5:1	16.5:1	10.5:1	1:1 (oxygen)	1:1.3 (oxygen)
Heating Value (BTU/lb)	18,000	19,916	21,480	?	?
SE at stoich AFR	?	?	?	?	?
Heat of Vapor	?	?	?	?	?
Boiling Point (C.)	-44F	-259F	?	?	?
RON	112	130	100+	?	?
MON	104	?	95+	?	?
Flame Speed (cm/sec)	39.0	33.8	?	?	?

Reston Publishing Co.

combustion chamber for burning.

2). Heating Value. How much heat energy is released by burning the fuel? The more heat released, the higher the potential power output from a fuel.

3). Specific Energy (SE). The SE of a fuel is obtained by dividing the fuel's heating value by its stoichiometric air-fuel ratio. This represents the amount of heat energy a fuel can deliver for a given amount of air drawn into the engine. In other words, how much heat energy a given amount of fuel can deliver when it burns in exactly as much air as necessary. A theoretical SE can be computed, or an SE can be obtained by observing the actual heat of combustion.

Specific energy ratio is a useful measure to compare fuels. It represents the ratio of the SE of a sample fuel to that of a reference fuel, usually iso-octane, which is similar to commercially available gasoline.

4). Chemical Expansion. This is the relationship between the volume of chemical reactants (the charge mixture prior to combustion) and the volume of chemical products of the combustion event (exhaust gases). The higher the ratio of products to reactants, the higher the combustion pressure and the higher the potential power output.

5). Air-Fuel Ratio Flammability Limits. Enriching air-fuel mixtures builds power until the added energy of the additional fuel is offset by decreasing flammability. Broad rich-flammability limits permit higher potential power.

6). Heat of Vaporization. Liquids consume energy as they boil, providing a cooling effect. This effect can significantly increase charge density, increasing the mass of air drawn into an engine. Fuels which consume a lot of heat to vaporize provide significant power benefits by effectively increasing engine volumetric efficiency.

7). Volatility. Volatility is determined by heating a fuel and determining the temperatures required to boil succeeding tenths of the fuel. A plot of the results yields a distillation curve. Ideally, a fuel should have a volatility range which enables it to be handled and transported in liquid form, yet

Energy Content

	BTU/lb	Density (lb/gal)	BTU/gal
Iso-octane	20.556	5.74	117,991
Toluene	18.245	7.20	131,290
#2 Diesel	18.310	7.09	129,800

Energy Content iso-octane, toluene, and #2 diesel. Aromatics like toluene are very dense due to the compact benzene ring structure. Not only are diesel engines more efficient than spark-ignition engines, particularly at part throttle, but diesel fuel has very high energy per gallon, enabling diesel-powered vehicles to produce very high mpg figures.

evaporate easily into a vapor in the intake system. Volatility is greatly affected by temperature and atmospheric and manifold pressure. This is why street gasolines are often formulated differently for summer versus winter, and why aircraft, which must operate under conditions of rapidly changing atmospheric pressure, often require special fuel with special volatility specifications. The volatility range or distillation curve of racing gas affects throttle response and engine startability.

8). Safety. The stability of a fuel should be such that it is not an explosion hazard during ordinary use or transport, and its toxicity to those handling it should not be beyond accepted limits for materials of similar composition. A good fuel should not blow up accidently or easily poison you.

9). Octane Rating. A fuel's octane rating characterizes the degree to which the fuel-air mixture will resist detonation—explosive spontaneous ignition of the remaining "end gases" as heat and pressure grows in the combustion chamber following normal ignition. Abnormal combustion can quickly destroy a high-performance engine under racing conditions. Motor octane number (MON), a measure of resistance to knock, is obtained at 900rpm in a single cylinder test engine with heated inlet air and advanced spark timing, and can be significantly lower than research octane number (RON), measured at 600rpm on a single cylinder test engine. Nevertheless, RON may occasionally be a better gauge of engine antiknock fuel requirements, particularly for street turbocharged

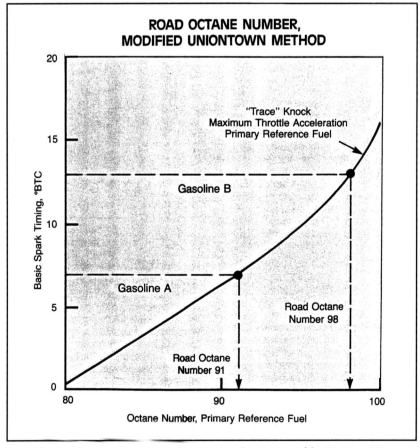

Road Octane Number compares the octane of fuels in multicylinder engines under real conditions. Octane number is obtained by comparing spark timing at trace knock to reference fuels. *Chevron*

Highly developed engines demand highly developed fuels, and breakthroughs in fuels tend to follow new engine developments. Intensive work goes into developing fuels optimized to produce every last horsepower possible within the rules in events like Formula One racing.

engines in which worst-condition octane requirement is only achieved rarely, under full boost conditions. The average of the two ((R+M)/2) is displayed by law on all USA street filling station pumps.

The octane of a fuel is defined as the percentage of iso-octane in a mixture of iso-octane and n-heptane that has equal resistance to abnormal combustion. One hundred percent iso-octane is called 100 octane; 100 percent n-heptane is zero octane. The reference fuel for octane ratings better than pure iso-octane is iso-octane plus 5 grams tetraethyl lead per gallon. Antiknock ratings of 110–120 (RON) and 105–110 (MON) are capable of meeting the most severe racing conditions in spark-ignition piston engines. Perhaps a better gauge for high-performance fuel requirements is road octane number (RdON), which is the antiknock performance of the fuel when subjected to real conditions found in multi-cylinder engines in automobiles under typical running conditions. RdON may be established on a chassis dyno or on a high-speed engine dyno configured to simulate racing conditions. In the Modified Uniontown Technique, fuel octane is evaluated in an engine operated at wide-open throttle across the rpm range during which spark advance is calibrated to produce a barely audible knock. Road octane number is calculated by comparing the spark advance of the test fuel to that of a primary reference fuel which has also been evaluated on the test engine. Given that road octane is often not available, the number racers and enthusiasts should be concerned about is the MON.

10). Price and Availability. The price of fuel varies from as little as $0.45 per gallon for propane, to over 100 times as much for nitromethane. A gallon of fuel varies in its range and power potential. Fuel availability varies widely. Gasoline is available on nearly any street corner; diesel is available at truck stops and some gas stations in all cities; propane is available in all cities for heating purposes and around the country for RV use; Natural gas is widely used in California, Texas, and some other states with high levels of air pollution, for fleets of trucks, buses and autos, and is widely used for heating and power generation; avgas is available at many airports. Racing gasoline is available in 55-gallon drums from specialty suppliers and distributors and some large oil companies and by the gallon at some race events. Methanol and nitro are available in 55-gallon drums with special handling requirements; Hydrazine, an explosive rocket fuel, is available only under limited circumstances for research.

11). Legality. Various racing sanctioning bodies have rules about what fuels are legal for various forms of racing, and testing methods that are designed to help prevent racers from cheating. The United States government, particularly the EPA, regulates the legality of various forms of gasoline and alternative fuels for street use. Fuels like propane and natural gas, which are usually stored under pressure, are subject to laws that specify construction standards for fuel tanks and their location on a vehicle.

12). Flame Speed. Clearly, some fuels burn faster than others and are more suitable for high-speed engines. Typically, high-speed, top-fuel drag cars shoot flames from the exhaust because there is no time for complete combustion. Fuel burning in the exhaust does not make power. Flame speed is dependent not only on the particular fuel, but on the air-fuel mixture strength and the swirl or turbulence in the combustion chamber, which has a much greater effect on flame speed than the particular variety, say, of gasoline.

13). Additives. Street gasoline and many other fuels use additives to improve some of the above characteristics. Several states now require so-called oxygenated gasolines for some or all of the year. Oxygenated gasolines contain oxygen-bearing alcohols or ethers that improve emissions by providing an additional source of oxygen already chemically bounded to the fuels which effectively leans mixtures in carburetted engines—and EFI cars at full throttle—for more complete combustion.

14). Specific Gravity. This measurement compares the weight of a fixed volume of fuel to that of the same volume of water. It is an indicator of fuel density.

15). Emissions Characteristics. A fuel with even one very high-performance characteristic (such as octane rating in propane) may be desirable even in the absence of the highest power potential. Stuffing more air and fuel in the engine with forced induction can produce as much power as necessary with a cheap, high-octane fuel like propane or natural gas while providing excellent emissions characteristics for street cars, and even possible exemption from emissions standards as a clean-burning alternative fuel.

Street, racing, and aviation gasolines, gasoline components like iso-octane, toluene and xylohexane, methanol, ethanol, nitromethane, nitropropane, hydrazine, propane, natural gas, diesel, and M85 all have at least some characteristics that can be considered "high performance" for the purposes of this book, although none of them is optimum for all circumstances.

2

Chemistry Lesson

Despite the uncountable number of different substances in the universe, there are only 230-odd different kinds of atoms. There are a few examples of substances made up of only one type of atom (pure metals, for instance), but most of the stuff we encounter every day—from water to whisky, from rubber to rat poison—is a combination of some of these atomic elements.

The most basic type of combination is a chemical compound in which the atoms of two or more different elements link with each other to produce a substance completely different from its constituent atoms. Water, for example, consists of two atoms of hydrogen chemically bonded to one atom of oxygen, hence its chemical symbol H_2O. The relative numbers of each atom that make up a compound is determined by the valence of the atoms—roughly, the number of electrochemical "hooks" they have. Note, though, that not all the "hooks" need be used. Hydrogen and oxygen can also combine in another way to produce a completely different material—hydrogen peroxide, H_2O_2.

This process by which elements join to form a compound, or by which a compound breaks down into its component atoms, is what is meant by a chemical reaction. The thing to understand about chemical reactions is that some are endothermic (they require net energy in order to occur) while others are exothermic (they generate net energy as they occur). Breaking bonds requires energy; forming bonds releases energy. Breaking weak bonds and forming strong bonds releases net energy. There is a strong tendency of chemical compounds to

TYPICAL HYDROCARBON CONFIGURATION

Hydrocarbons are classified as paraffins, olefins, naphthenes, and aromatics. Olefins, with double bonds, tend to have higher energy but lower octane and lower stability. They are only used rarely in racing fuel blends where high energy content is vital. Although aromatics are illustrated with conjugated double bonds, in reality, they are highly stable and high octane because the conjugated bonds resonate and behave like one-and-a-half bonds. *Lubrizol*

seek out the state of greatest stability, i.e. the state with the least energy. Nitroglycerin, for example, is so unstable (high energy) that merely shaking it can cause a powerfully explosive chemical reaction that yields heat, large amounts of product volume, plus lower energy compounds of greater stability.

Fuels produce energy by the breaking and forming of chemical bonds to build more stable products. Often, but not always, this involves oxidation (combining with oxygen). Fuels like nitromethane can exothermically decompose to form lower energy products without any oxygen at all, although they liberate even more energy with oxygen. Explosives like nitroglycerin decompose in an explosive burst of energy entirely in the absence of gaseous oxygen.

Hydrocarbons

Hydrogen, a single-bonding atom (valence of 1), and carbon, a four-bonding (quadravalent) atom, can combine in many thousands of different molecular structures called hydrocarbons. Hydrocarbons, when oxidized, liberate huge amounts of energy.

The valence of an atom is based on the number of electrons in the outer shell(s). Atoms with partially filled shells are not stable, but by sharing electrons with other atoms (covalent bonding, the type of bonding in most organic molecules), or having electrons donated by another atom (ionic bonding), the result is a stable chemical compound. Oxygen (a two-bonding atom), hydrogen, carbon, nitrogen (valence of 3 or 5), and many other atoms are not stable at ambient temperatures, which is why oxygen in the air exists as O_2 molecules, that is, two atoms bonded together with twin covalent bonds. Carbon can similarly combine with itself in single, double, or triple bonds. Two carbon atoms form single-bond hydrocarbons with six hydrogen atoms (ethane), double-bonded hydrocarbons with four hydrogen atoms (ethene or ethylene), or triple-bonded with two hydrogen atoms (ethyne or acetylene).

Hydrocarbons with only single bonds are referred to as saturated (saturated with the maximum amount of hydrogen), while those with some double or triple bonds are called unsaturated. The single-bonded or saturated hydrocarbons are more stable than double- or triple-bonded unsaturated hydrocarbons, since compounds like oxygen can easily react with them across the double or triple bond. Unsaturated hydrocarbons are commonly found in low-performance automotive fuels (although not usually in high-performance racing or aviation fuels), but triple-bonded hydrocarbons are not commonly present.

The simplest hydrocarbon is methane, or CH_4, with one carbon atom bonded to four hydrogen atoms. Gasolines, propane, natural gas (mostly methane), diesel fuel, and fuel oil are all hydrocarbons. Gasoline, with the highest heating value of any of the normal racing fuels, is a mixture of scores or hundreds of hydrocarbons. Various hydrocarbons can have radically different performance characteristics even with slightly different molecular structures.

There are four major classes of hydrocarbon compounds:
alkanes/paraffins ($C_nH_{(2n+2)}$),
cycloparaffins/naphthenes ($C_nH_{(2n)}$),
alkenes/olefins ($C_nH_{(2n)}$), and
aromatics ($C_nH_{(2n-6)}$).

Paraffins

Paraffins (also called alkanes) are a series of saturated hydrocarbons of which methane is the simplest member, followed by ethane, with two carbon atoms, then propane with three, and so on. The carbon atoms can be arranged in a straight line, or in branched chains. As the number of hydrogen atoms increases, there are increasing numbers of different chemical structural arrangements of the same number of carbon and hydrogen compounds. Alternate arrangements of the same basic chemical formula are called isomers, and they are entirely separate compounds with their own individual boiling point, anti-knock index (AKI), and so on. For example, iso-octane, a branched

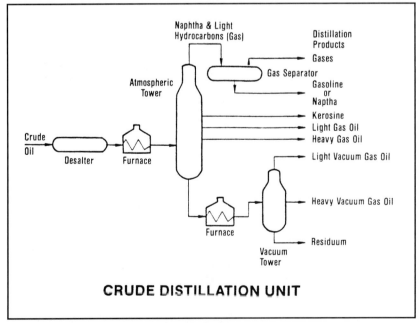

Crude distillation unit separates hydrocarbons by boiling point, which is highly corrolated to molecular weight. *Lubrizol*

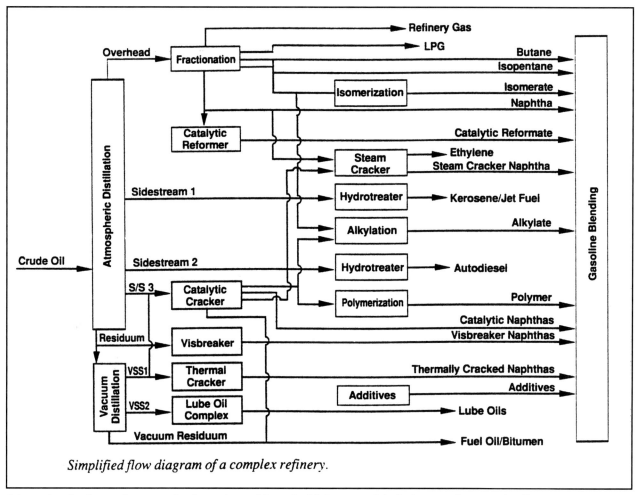

Simplified flow diagram of a complex refinery.

Schematic of refinery shows methods used to achieve the highest possible fuel output from crude oils. In some cases, severe methods break down large molecules to gasoline or diesel size while polymerizations are used to build up very small molecules into fuel-sized compounds. *SAE*

The Fractions Obtained from the Distillation of Petroleum

Percentage of Total Volume	Boiling Point (°C)	Approximate Number of Carbon Atoms	Common Names of Products
1–2%	<30	C_1–C_4	Light hydrocarbons (methane, ethane, propane, butane)
15–30%	30–200	C_4–C_{12}	Naphtha or straight-run gasoline
5–20%	200–300	C_{12}–C_{15}	Kerosene
10–40%	300–400	C_{15}–C_{25}	Gas oil
Undistilled	>400	>C_{25}	Residual oil, paraffin, asphalt

Distillation fractions of hydrocarbons.

paraffin, has an octane rating of 100 and a boiling point of 99 degrees C, while n-octane, a straight-line paraffin, has a blending AKI of -17 and a boiling point of 126 degrees C. Besides iso-octane and n-octane, there are 16 other isomers of octane. In fact, the more carbon atoms present in a hydrocarbon, the more possible isomers there are. Only the simplest paraffins, methane and ethane, do not have isomers.

The group CH_3, derived from methane, is called a methyl group or methyl radical, or more generally an alkyl group. It is often found attached to carbon atoms as a building block of more complex isomeric hydrocarbons, as are other more complex straight line hydrocarbon alkyl groups: The ethyl group (C_2H_5), the propyl group (C_3H_7), the butyl group (C_4H_9), and so on. Each of these has "group isomers," meaning there are two propyl alkyl groups (isopropyl and n-propyl), three butyl alkyl groups (n-butyl, isobutyl, and tertiary butyl), and so forth. Isobutane can be thought of as propane (CH_3-CH_2-CH_3) with one of the hydrogen atoms bonded to the middle carbon atom replaced by a methyl group, which is why isobutane is also referred to as 2-methylpropane. Iso-octane can also be referred to as 2, 2 , 4-trimethylpentane.

Naphthenes

Naphthenes or cycloparaffins are, in simplest form, CH_2 groups arranged in a circle. Cyclopropane, the simplest, is three carbon atoms bonded together in a circle, each with two hydrogen bonds. The most stable, with minimal distortion of the carbon bond angles, have five carbon atoms (cyclopentane) or six (cyclohexane).

Olefins

Olefins have the same carbon-to-hydrogen ratio as naphthenes, but olefinic hydrocarbons have entirely different characteristics and behavior. The double bond is very reactive and high energy, all things being equal, giving olefin-based gasolines higher energy than gasolines made of single-bond hydrocarbons. Essentially, they are unsaturated straight-line or branched hydrocarbons; they have at least one double carbon bond. The name of various olefinic isomers changes according to the position of the double bond. If the double bond is between the first and second carbon atoms, for example, CH_2=$CHCH_2CH_2CH_3$ it is called pentene-1 or 1-pentene, or even pent-1-tene.

More than one double bond can occur, the simplest of which is butadiene (CH_2=CH-CH=CH_2), and these are very undesirable in automotive fuels because they are extremely prone to sudden oxidation. The maximum number of double bonds occurs where every other carbon bond is double. These are called conjugated double bonds, and they yield the worst type of engine fuel due to their extremely high reactivity.

Like paraffins, olefinic groups or radicals can be combined as building blocks of more complex hydrocarbons. Olefinic groups include the vinyl group (CH_2=CH-) and the alkyl group (CH_2=CH-CH_2-).

Aromatics

Originally, this term applied to certain hydrocarbons with a pleasant "aromatic" odor. It now applies to a class of hydrocarbons based on a unique six-membered ring which at first appears to have alternating (conjugated) double carbon bonds, but the bonds actually resonate between single and double bondedness, as if the double bonds were not in fixed positions. Aromatics are consequently not particularly reactive like normal unsaturated hydrocarbons. The simplest aromatic is benzene (C_6H_6).

Various radicals or hydrocarbon groups can substitute for hydrogen atoms bonded to the hexagonal carbon ring of the aromatic structure. Toluene, an extremely high octane blending agent for racing or aviation gasoline, is basically benzene with a methyl group substituted for one hydrogen atom and can be referred to as methylbenzene. Xylene, or dimethylbenzene, has two methyl groups. Both these compounds have extremely high octane and are commonly found in racing gasoline. They are also very good solvents and can be commonly bought at hardware stores by the gallon.

Aromatic rings can be fused together in double or triple rings (naphthalene and anthracene) or even more complex polynuclear aromatic structures (PNAs) with or without auxiliary radicals. These are sometimes referred to as polycyclic aromatic hydrocarbons. Some of the complex PNAs are highly carcinogenic and are found in small amounts in automotive exhaust gases.

Other Hydrocarbons

Terms like alkylates, catalytic reformates, isomers and so on are frequently used to describe components of gasolines and other fuels. In some cases these are terms for branched hydrocarbons based upon the methods used to produce them.

Hydrocarbon Combustion

All hydrocarbons can be ignited in the presence of oxygen to form carbon dioxide and water. Since, practically speaking, all piston engine combustion takes place in air rather than pure oxygen, the stoichiometric combustion of iso-octane (representative of racing gasoline) is expressed as:
$C_8H_{18} + 12.5O_2 + 47N_2 \rightarrow 8CO_2 + 9H_2O + 47N_2$

Carbon has a molecular weight of 12, hydrogen of 1, and oxygen a weight of 16. Therefore, 114 grams of iso-octane reacts with 400 grams of oxygen (or roughly 1,715 grams of air) to form 352 grams of carbon dioxide and 162 grams of water. The stoichiometric (chemically perfect) air-iso-octane ratio for complete combustion is roughly 15:1. The brew of hydrocarbons in actual gasoline typically has a stoichiometric air-fuel ratio of 14.6:1. Because air is used to oxidize the hydrocarbons in gasoline, at the high temperatures of combustion (as high as 1,500 degrees), small amounts of nitrogen combine with oxygen to form oxides of nitrogen, referred to as NO_x, since the relative proportions of nitrogen and oxygen are unpredictable, and there are three common oxides of nitrogen. All are considered pollutants. When there is too little oxygen for stoichiometric combustion, another pollutant, carbon monoxide, will be formed. Too much oxygen can result in less carbon monoxide but more NO_x. What's more, in a real engine, some unburned hydrocarbons will survive the process unburned, and these may well contain carcinogenic PNAs.

The simpler hydrocarbon propane (a major component of LP gas) combusts according to the following:

$C_3H_8 + 5O_2 + 18.8N_2 \rightarrow 3CO_2 + 4H_2O + 18.8N_2$

Natural gas, mostly methane, combusts according to the following formula:

$CH_4 + 2O_2 + 7.5N_2 \rightarrow CO_2 + H_2O + 7.5N_2$

Both these fuels are considered clean-burning. Diesel fuels are heavier hydrocarbons than gasolines with boiling points over 400 degrees.

Monohydric Alcohols

Alcohols are, essentially, partially oxidized hydrocarbons, and include one atom of oxygen in a hydroxol group (OH-). They are of the general formula $C_nH(2_n)OH$. Methanol (methyl alcohol), ethanol (ethyl alcohol), propanol (propyl alcohol), and butanol (butyl alcohol) have all been used as automotive fuels. Alcohols with more than one OH group are called polyhydric.

Alcohol fuels have only about half the heat energy of gasoline, since they are already partially oxidized. However, they actually have a slightly higher SE because they need much less air to burn—a given amount of air can burn much more alcohol than gasoline—and because their wide flammability limits allow mixtures extremely rich of stoichiometric. Alcohols are sometimes called oxygenated fuels, and their presence in combustion (often mixed with gasoline or other fuels) tends to reduce emissions by contributing chemically bonded oxygen to the process. Alcohols have a high heat of vaporization, providing a substantial cooling effect to the inlet air, which significantly increases air density and therefore power. They also have very high resistance to detonation.

In common monohydric alcohols, the OH- group replaces one of the hydrogen atoms in a paraffin, and its polarity gives alcohols certain properties which distinguish them from hydrocarbons, such as solubility in and affinity with water.

As the number of carbon atoms in alcohols increase, the distinctive properties diminish. Methanol,

Names and Formulas for Some Common Hydrocarbons

Formula	Name
CH_4	Methane
C_2H_6	Ethane
C_3H_8	Propane
C_4H_{10}	Butane
C_5H_{12}	Pentane
C_6H_{14}	Hexane
C_7H_{16}	Heptane
C_8H_{18}	Octane

Uses of the Various Petroleum Fractions

Petroleum Fraction in Terms of Numbers of Carbon Atoms	Major Uses
C_5–C_{10}	Gasoline
C_{10}–C_{18}	Kerosene
	Jet fuel
C_{15}–C_{25}	Diesel fuel
	Heating oil
	Lubricating oil
$>C_{25}$	Asphalt

Names and uses of common hydrocarbons.

with one carbon atom, has the highest percentage of oxygen by weight and is most distinctive in electrical and corrosive properties, and affinity for water; ethanol follows. There are two isomers of propanol: n-propanol (CH_3-CH_2-CH_2-OH) and isopropyl (rubbing) alcohol (CH_3-CH-OH-CH_3). There are three isomers of butanol. We are mainly concerned with methanol and ethanol.

The stoichiometric combustion of methanol is expressed as follows:

$CH_3OH + 1.5O_2 + 5.64N_2$ -> $CO_2 + 2H_2O + 5.64N_2$

The stoichiometric combustion of ethanol is expressed as follows:

CH_3-CH_2-OH + $3O_2$ + $11.28N_2$ -> $2CO_2 + 3H_2O + 11.28N_2$

Alkyl Ethers

These are isomers of monohydric alcohols but contain a single oxygen atom linked to two alkyl groups (not necessarily of the same type) rather than an alkyl group and a hydrogen atom. The simplest ethers (dimethyl ether and ethyl methyl ether) are too volatile for use in liquid automotive fuels. The ones commonly combined with gasolines are methyl tertiary butyl ether (MTBE), written as CH_3-O-C(CH_3)$_3$, ethyl tertiary butyl ether (ETBE), written as C_2H_5-O-(CH_3)$_3$, and tertiary amyl methyl ether (TAME), in which one of the methyl groups in the tertiary butyl structure of MTBE is replaced by an ethyl group.

The stoichiometric combustion of MTBE is expressed as follows:

CH_3-O-C(CH_3)$_3$ + $2.5O_2$ + $9.4N_2$ -> $4CO_2 + 6H_2O + 9.4N_2$

Nitroparaffins and Other Nitro-group Compounds

The simplest and most common nitro-group fuel is nitromethane—"nitro." An oily, clear liquid, it is used in making dyes and resins, in organic synthesis, and as a fuel for rockets... and top-fuel dragsters. Written as (CH_3NO_2), it includes twice the oxygen of monohydric alcohol, as well as nitrogen—essentially, a nitro group (NO_2) substitutes for a hydrogen atom on a paraffin.

Nitromethane is a monopropellant. It can actually combust without oxygen, which makes it useful as a fuel, but dangerous. As a monopropellant, its combustion yields water, carbon dioxide, carbon monoxide, hydrogen, and nitrogen. Stoichiometric burning of nitromethane yields water, CO, and gaseous nitrogen. It has a low heating value, but a specific energy over twice that of gasoline, due to the high energy nitrogen bonds formed during combustion, and because you can burn it in air-fuel ratios as high as 1:1. A naturally aspirated engine burning nitro versus gasoline has the potential to make over double the power.

The reaction of nitromethane as a monopropellant may be expressed as:

CH_3NO_2 -> $0.25CO_2 + 0.75CO + 0.75H_2O + 0.75H_2 + 0.5N_2$

Nitromethane's stoichiometric combustion in air is expressed as:

CH_3NO_2 + $1.5O_2$ + $2.82N_2$ -> $CO_2 + 1.5H_2O + 2.82N_2$

Nitroethane ($C_2H_5NO_2$) and nitropropane ($C_3H_7NO_2$) have similar properties. Nitropropane will mix with gasoline, while nitromethane will not (mixes with methanol).

Chemistry of Monopropellants (Hydrazine, Thermoline, etc.)

Hydrazine (H_2N-NH_2) is a colorless, fuming, corrosive hygroscopic liquid used in jet and rocket fuels. (Hygroscopic means it absorbs water from the atmosphere.) Hydrazine boils at 113 degrees C. When mixed with nitromethane, it forms explosive salts (hydrazinium salt of methazanic acid) that require only the oxygen in nitromethane to combust.

3

Lubricants

The first steam engines used natural fats and oils (tallow, lard, olive oil, etc.) for lubrication. But starting around 1850, natural fats gradually began to be replaced by mineral oil refined or separated from crude petroleum. From 1870 to 1920 there was a steady improvement in mineral oil due to new refining techniques such as de-waxing. In 1911, the Society of Automotive Engineers (SAE) designated the standard system for measuring oil viscosity, with six winter grades (0W-25W) for oils designed to flow well in cold weather, and five grades (20-60) for oils flow-rated at 212 degrees. In 1929, the first synthetic lubricant, formed by combining simple light hydrocarbons into uniform heavier molecules of "synthetic" oil, was produced in a pilot plant, but the scheme was soon abandoned for economic reasons.

Early internal combustion piston engines used "straight" mineral oil, with no additives. In those early days, engines were lightly stressed, and oil change intervals were short—typically 1,000 to 3,000 miles. In spite of the frequent oil changes, engines did not last very long. Pistons, rings, bearings, and other high-wear parts wore out quickly, and even well-maintained engines were found to be full of carbon, sludge, varnish, and other components of oil breakdown and contamination. Trade magazines of the 1930s, 1940s, and 1950s were full of ads for replacement parts and technical articles detailing techniques for cleaning clogged oil passages.

It was not until the 1930s that original equipment and aftermarket oil filters began appearing on passenger cars in any volume. In 1940, the U.S. Army first required oil filters on all its vehicles. The Fram spin-on filter was introduced in 1956.

According to Chevron, the average passenger car engine lasted only 45,000 miles in the 1950s. By contrast, in the 1990s, advanced engines equipped with computer-controlled port fuel injection and advanced lubricants typically lasted four times as long before a rebuild was required. In fact, test engines lubricated with the best synthetics showed negligible wear at 200,000 miles!

In the meantime, due to the Wehrmacht's shortage of mineral oil and problems of severe cold on the Russian front in WWII, German chemists had provided rapid development of synthetic oils which had markedly superior performance in extremely cold or hot weather, and that could be produced from coal or natural gas.

Following the war, synthetic lubricants underwent additional rapid development for jet engines and aircraft. Synthetics were first required in the wheel bearings of early-1960s carrier-based aircraft. Many other synthetics were developed for special aviation use. Today, synthetics are required in all jet engines.

Around 1976, Mobil introduced the first retail synthetic oil available in the U.S. for street automotive use. (Synthetics had come to Europe a few years earlier.) Other companies followed suit in the 1980s and 1990s, and most major brands now offer synthetic oils. Still, less than 1 percent of automotive lubricants are synthetic. In the meantime, ordinary refined mineral oils have improved due to development of superior refining techniques such as wax reformation.

How is it that modern engines last so long compared to engines of the past? Electronic fuel injection has eliminated the cylinder wall washing that used to occur under cold start in carbureted engines. Higher operating temperatures, better PCV systems, and the switch to unleaded gasoline have helped eliminate corrosive acids in the crankcase. Materials and machining have also improved, but much of the credit for longer-lasting engines goes to improved motor oils. American Petroleum Institute (API) service categories have progressed from the SA and SB oil standards typical of the 1950s to the much-improved SG or SH standards and include synthetics which can stretch engine life to a quarter million miles or more.

Understanding Motor Oil

Crude oil, as it is pumped out of the ground, contains hundreds of kinds of hydrocarbons from thick, tar-like asphalts to thin, volatile liquids like benzene or heptane, or even light, colorless gases like methane or propane. Refiners separate crude oil into scores of useful compounds. The base stock for motor oil is made from medium-size hydrocarbon molecules with 25 to 40 carbon molecules which are neither too thick nor too thin, and can be pumped under high pressure between moving metal surfaces of an engine to provide a slippery lubricating film that prevents metal-to-metal contact and all but eliminates friction. Motor oil hydrocarbons, like gasoline, consist of olefins, paraffins, naphthenes, and aromatics. Simple aromatics—which make good gasoline components due to their ability to resist detonation—also make good motor oils due to their thermal stability. Polynuclear aromatics do not make good oils.

The Refining of Lubricants

Crude oil, formed over eons from the remains of tiny prehistoric plants and animals, contains a complex stew of hydrocarbons ranging from methane, with one carbon atom, to compounds with 50 or more carbon atoms, plus impurities like sulfur. Mineral lubricating oil is refined in the same process which produces other oil products, including gasoline. Because the boiling range increases with chemical weight (the number of carbon atoms), all but the heaviest products such as asphalt can be separated by distillation. Refining begins by de-salting; the crude is then vaporized into a fraction column which separates the lighter compounds such as gasoline according to boiling range. Oils (including lubrication oils), which vaporize above 650 degrees F are then distilled by vacuum fractioning at less than atmospheric pressure, to lower their boiling points and so prevent decomposition at the over 700 degree temperatures which would otherwise be required. Given the complexity of crude oil (a 25 carbon atom paraffin with up to 52 hydrogen atoms has 37,000,000 different chemical arrangements), and substantial differences between crude from various parts of the world, refiners are careful about crude selection, isolation of similar fractions with similar boiling range, subtractive processing to remove undesirable contaminants, and final blending of finished stocks.

Following vacuum fractioning, solvent extraction separates aromatic compounds by mixing the various distillates with a solvent such as furfural, allowing the mixture to settle and removing the solvent. Two phases form: an extract phase rich in aromatics, and a rafinate phase rich in paraffinic hydrocarbons. Wax is then removed by mixing in methyl ethyl ketone, then cooling to 10 to 20 degrees F, at which time wax crystals can be filtered from the oil. Optional hydrofinishing passes heated oil and hydrogen over a catalyst to remove color bodies and unstable compounds such as nitrogen and sulfur. Alternately, lube oils may be formed by hydrocracking, in which the molecular structure of many feedstock compounds is changed; aromatics become naphthenes, naphthene rings are broken, and paraffins are rearranged or broken up. A significant proportion of the feedstock is converted to lower boiling point compounds.

The Functions of an Engine Lubricant
Reduction of Friction

The main purpose of motor oil is to reduce friction by preventing direct metal-to-metal contact between moving parts. This is referred to by chemists as thick film lubrication, even though the oil films are sometime only .000025in thick—1/50th the thickness of a human hair!

Friction is a force that resists relative motion between two contacting bodies. The coefficient of friction is the ratio of the frictional force to the perpendicular load between the contacting surfaces. Unlubricated metal surfaces of ordinary finish typically have a coefficient of friction of 1. Clean metal surfaces in a vacuum might have a coefficient of friction of 100–200 or more. Well-designed and well-lubricated mechanical systems typically have a coefficient of friction between 0.005 and 0.000005.

Even precisely polished metal surfaces have microscopic imperfections which will create heat and friction if allowed to rub together.

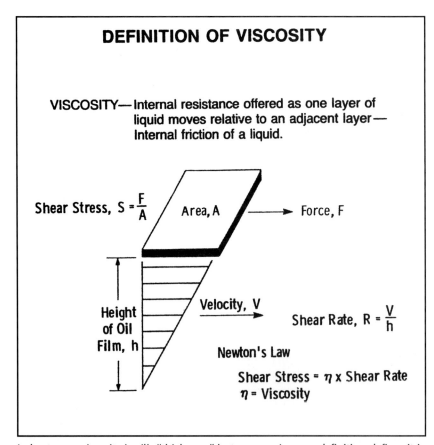

In lay terms, viscosity is oil's "thickness," but a more rigorous definition defines it in terms of internal resistance as one layer of a liquid moves relative to others. *Chevron*

ENGINE LUBRICATING OIL ADDITIVES

TYPE	REASON FOR USE
Viscosity Index Improvers	Reduce viscosity change with temperature; reduce fuel consumption; maintain low oil consumption; and allow easy cold starting.
Detergents, Dispersants	Keep sludge, carbon, and other deposit-forming material suspended in the oil for removal from the engine with drains.
Alkaline Compounds	Neutralize acids. Prevent corrosion from acid attack.
Antiwear, Friction Modifiers	Form protective films on engine parts. Reduce wear, prevent galling and seizing. Reduce fuel consumption.
Oxidation Inhibitors	Prevent or control oxidation of oil, formation of varnish, sludge, and corrosive organic compounds. Limit viscosity increase that occurs as oil mileage increases.
Rust Inhibitors	Prevent rust on metal surfaces by forming protective surface films or by neutralizing acids.
Pour Point Depressants	Lower "freezing" point of oil, assuring free flow at low temperatures.
Antifoam Agents	Reduce foam in the crankcase.

Engine lubricating oil additives. *Chevron*

Metal-metal contact of moving parts can create enough heat to weld the high spots together, which then tear apart and re-weld, and so on. If this "scuffing" continues, engine parts will eventually weld together with enough strength to seize the engine.

Although pistons and rings are lubricated just by oil splash, oil is pumped under pressure to bearings and other high-stress parts, which results in a continuous wedge of oil being formed between the moving surfaces, a situation referred to as hydrodynamic lubrication. Mechanical pumping replenishes the oil film between, for example, a metal bearing and a rotating journal, so an equilibrium thickness of oil film is established according to the ratio of oil input and oil leakage at the bearing ends.

The film thickness is reduced with increasing load, increasing oil temperature, which increases leakage, by changing to a lower viscosity oil, or by reducing the journal speed. Viscosity not only regulates the sealing effect of oils but also the rate of oil consumption and heat generation from internal friction. Hydrodynamic theory assumes that friction occurs only in the oil film separating moving parts, and is determined mainly by the viscosity of the oil. A hydrodynamic film of oil under high pressure between two metal surfaces can literally be more rigid than the metal surfaces it is lubricating, under which conditions additional loading has the effect of deforming the metal surfaces!

In the real world, the simplifying assumptions underlying hydrodynamic lubrication are seldom valid. Shock loading, steady heavy loading, high temperatures, slow speed or critically low viscosity can prevent hydrodynamic lubrication and allow intermittent contact between lubricated surfaces. Temperatures soar, and the contacting surfaces are progressively destroyed. This is why antiwear additives are now added to oil to form organic films that coat metal and prevent its destruction in case the surfaces actually do touch.

Heat Removal

An important function of motor oil is to transfer heat to an oil cooler or into the engine's cooling system. The lubricant must maintain its integrity under high and changing heat conditions. In particular, it must not oxidize or undergo other thermal changes which affect its ability to reach areas where cooling is essential. In order to meet these requirements, additives are normally required.

Containment of Contaminants

Lubricants must continue to perform well even in the presence of contaminants such as water, particulate matter, and blow-by combustion products which make their way into the lubricant. In the case of engines, oil filters remove larger particulate matter, although the most wear damage is from particles 10–20 microns in size. Other crud too small to be filtered must be held suspended in the oil until it is changed.

Other Required Properties of a Lubricant
Low volatility under all operating conditions
This is a characteristic of the oil base stock and cannot be improved with additives.

Proper flow characteristics at the target operating temperature range
Again, the characteristics of the base stock establish basic flow characteristics, but pour-point depressants and viscosity improver additives can improve flow performance.

Sufficient stability to maintain specification characteristics for reasonable time
It is the additives, not the base stock, which largely determine the stability of the lubricant. Additives have made a major contribution toward improving oil's ability to remain stable in the face of extreme temperatures and contamination by water, unburned fuel, and harmful acidic combustion products.

Compatibility with other system materials
The additives in a lubricant should not damage gaskets, rubber, plastic or vinyl seals, clutch plates, or any metallic components of the system, including bearings.

Additives
Additives include metallic, ashless and polymeric dispersants, corrosion inhibitors, antiwear additives, viscosity improvers, and pour-point depressants. They are vital to the performance of modern oils.

Additives are expensive, and they are in oil for a reason, each additive performing a specific vital task. The right balance of additives is essential, since too much of any one can throw off the balance of the package of additives and produce harmful side effects that are not necessarily predictable. The results of complete additive packages are determined by exhaustive testing in actual engines. For these reasons, and because in any case, extra additives cannot revitalize worn-out oil. Racers and performance enthusiasts should select oils with characteristics that meet their requirements, as delivered from the manufacturer. The following discussion of additives proceeds with this in mind.

Antiwear Additives
In addition to its primary thick-film lubricating function, motor oil controls engine deposits, prevents corrosion, and provides thin-film lubrication in areas of the engine where thick films can't form, because the relative speeds of moving parts are too slow, and/or because the loads are too high. In these cases, metal-to-metal contact may be inevitable. To prevent catastrophic wear and metal-metal welding, oil additives like zinc dithiophosphates (the active ingredient in engine assembly lubes) are used. Where metal surfaces touch and temperatures climb, the additives form a thin inorganic film which shears more easily than the base metal, allowing the parts to slide against each other without causing catastrophic failure. The antiwear additives do not work well in cold oil, which is one reason starting a cold engine causes a lot of wear, and they eventually get used up, which is one reason oil must be changed periodically.

Abrasive wear is caused mainly by foreign particulate matter in the oil, and it is essential to keep oil well filtered and prevent particles from entering the engine via intake air. Corrosion wear is caused by acidic combustion products that are forced past the rings. Alkaline phenate and sulfonate additives control corrosion wear by neutralizing the acids.

Antiwear Friction Modifiers
In the eternal quest for improved fuel economy, friction modifiers have become increasingly important. Seventy percent of lubrication-related friction is caused by oil molecules rubbing against each other; lower viscosity oils reduce this friction. The other 30 percent of friction comes from parts protected by thin-film lubrication rubbing against each other. Friction modifiers further smooth this contact, providing a small additional reduction of friction.

Detergents and Dispersants
A vital task of any modern oil is to keep sludge, varnish, and carbon from clogging up the works of an engine. Detergents and dispersants are added to the oil to keep suspended small bits of trash too small to be filtered out, so they circulate with the oil until it's changed. If the detergents and dispersants wear out, or so much crud accumulates that the oil can no longer hold it, sludge and varnish will soon coat the internal surfaces of an engine.

Metallic detergents are really dispersants; they keep dirt and debris dispersed, rather than actually dissolving it. Metallic dispersants have a long hydrocarbon tail and a polar group head. The oleophilic hydrocarbon tail makes the molecule soluble in the oil, while the polar head attracts particulate contaminants.

Ashless dispersants used for multiviscosity oils include high molecular weight polymeric dispersants; lower molecular weight additives are used for situations in which viscosity improvement is not necessary. Both are more effective than metallic dispersants in controlling sludge and varnish deposits which could form during intermittent or low-temperature operation. Like the metallic dispersants, ashless dispersants contain a polar head group containing oxygen and/or nitrogen and/or phosphorus, attached to a high molecular weight hydrocarbon tail.

Polymeric ashless dispersants are long chain molecules in which a hydrogen or alkyl group is connected via a carbon molecule to an oleophilic hydrocarbon tail, and to the next identical link in the polymer chain via a CH_2 group.

Anticorrosion Additives

Leaded fuels high in sulfur were once major sources of acids in motor oil, and the elimination of lead and sulfur have made things easier on motor oils. Nevertheless, nitrogen-based acids, water, and organic peroxides still make anticorrosion additives an important part of modern oils.

Oxidation inhibitors are designed to prevent lubricant breakdown from oxygen attack on the base fluid. This occurs either by destruction of free radicals by chain breaking or by interaction with peroxides involved in the oxidation mechanism. Phenolic antioxidants are chain-breakers while zinc dithiophosphates are peroxide destroyers.

Bearing materials corrode due to reactions between the oxides of the bearing materials and acids resulting from blow-by gases or from oxidation of the base lubricant itself. Oxidation inhibitors reduce corrosion by neutralizing acids. Additionally, dithiophosphates form a protective film on bearing surfaces which makes them impervious to acidic attack, and at the same time provides thin-film anti-wear protection.

Viscosity Improvers

Almost all widely available oils for gasoline engines now meet API SG (or the newer SH) service ratings, which indicate that all have acceptable levels of antiwear additives, detergents and dispersants, and so forth. While special oils are available for aviation use, for alcohol, nitromethane, diesel, and propane and natural gas fuels, and while most are available as both mineral oil and synthetic oil, the main criteria for selecting oil has to do with the oil's viscosity. Viscosity measures how thick the lubricant is at a given temperature. Initially, all oils had a single viscosity, with six grades from 0W to 25W rated by the SAE for use at low temperatures, and five more grades from 20 to 60 for summer use.

The early single viscosity oils had the serious drawback that the thicker summer oils wouldn't flow at low temperatures, causing hard starting and rapid wear during cold start conditions. Winter oils were fine for starting, but could thin out so much in a hot engine (even in the winter) that oil consumption was a problem, and sometimes disastrous metal-metal contact occurred.

Chemists created multigrade oils by adding viscosity index (VI) improver additives to single-weight oils. These long-chain polymers allow the use of thinner single-weight base stocks for good low-temperature pour performance, but cause the base oils to perform like a thicker single-weight oil at the 212 degree test point. Virtually all car makers rec-

SAE J300 JUN 87
VISCOSITY GRADES FOR ENGINE OILS

SAE Viscosity Grade	Viscosity[1] (cP) at Temperature, °C, Max.	Borderline Pumping Temperature[2], °C, Max.	Viscosity[3,4] cSt at 100°C, Min.	Max.
0W	3250 at −30	−35	3.8	—
5W	3500 at −25	−30	3.8	—
10W	3500 at −20	−25	4.1	—
15W	3500 at −15	−20	5.6	—
20W	4500 at −10	−15	5.6	—
25W	6000 at − 5	−10	9.3	—
20	—	—	5.6	< 9.3
30	—	—	9.3	<12.5
40	—	—	12.5	<16.3
50	—	—	16.3	<21.9
60	—	—	21.9	<26.1

Note: 1 cP = 1 mPa·s; 1 cSt = 1 mm^2/s.

[1] ASTM D 2602 Cold Cranking Simulator.
[2] ASTM D 3829 or CEC L-32-T-82 for SAE 0W, 20W, and 25W; ASTM D 4684 for SAE 5W, 10W, and 15W.
[3] ASTM D 445.
[4] Some engine manufacturers also recommend limits on viscosity measured at 150°C and $10^6 S^{-1}$.

Viscosity-temperature characteristics of motor oils. Viscosity grades differ relative to temperature. *Chevron*

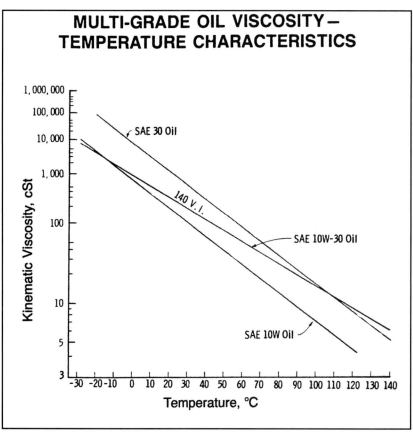

Multigrade oil viscosity-temperature characteristics. *Chevron*

ommend multiviscosity oils in new cars today, though not generally the thicker 10W-40 or 20W-50 oils. New cars perform well (and get better fuel economy) on API SH grade 5W-30 or 10W-30 oil, thereby avoiding the possibility of ring sticking and other problems that can occur in oils loaded with VI improvers. Those thick, honey-like oil additives—essentially concentrated VI improver in a can—can cause similar problems. Engineers now know that for street-driven normally aspirated cars, thin oil's low temperature flow characteristics help avoid significant wear at start-up time, which is much more important than a little less oil pressure in a hot engine. This is particularly true in overhead cam engines, where the cam bearing is located far from the oil pump. Oils designed to meet Energy Conserving and Energy Conserving II oil standards are all lightweight oils. The latter typically reduce friction 2.7 percent over traditional oils.

VI improvers are usually oil-soluble organic polymers with molecular weights in the 10,000 to 1,000,000 range. The polymer molecule is swollen by the oil, so the higher the temperature, the more the polymer swells, and the greater the thickening effect. This counteracts the natural tendency of oils to thin as they heat up. However, the performance of the VI improver is also affected by its resistance to mechanical shear, which decreases with increases in molecular weight, and so tends to cause a loss of viscosity. The ultimate performance of the lubricant is determined by this interplay, in which an economic balance must be established between these two tendencies.

Pour-point depressants are added to motor oil to prevent the oil from congealing at low temperatures, which is associated with paraffin wax components of mineral oil crystallizing. It is not economically feasible to remove all wax from mineral oil, nor is it desirable from a lubricative point of view. Pour-point depressants cannot prevent the formation of wax, but help prevent oil molecules from adhering to the wax crystals. Instead of oil "clumping" on the crystals, the oil remains separate and continues to flow.

Factory turbocharged engines pump high-pressure oil through the hot center section of a turbo, to help cool the turbo bearing, which can really heat up engine oil. So-called "turbo" oils are multiviscosity oils with 50 on the right side of the equation—anywhere from 5W-50 to 20W-50. Builders of both normally aspirated and forced-induction race engines do not have the same concerns about cold start performance, so the oils are always high viscosity, but street turbo engine oils must both prevent low temperature wear and stand up to destructive levels of heat under high loading in very hot weather. These cars typically have oil coolers, but when it comes to oil in such engines, it is hard to justify anything but synthetic.

Aftermarket Oil Additives

Oil additives have been around forever. Some years back, the heavy advertising came from STP. Andy Granatelli used to claim that STP would reduce engine wear and improve gas mileage; eventually, the Federal Trade Commission restricted STP from making such wild and unsubstantiated claims. The latest oil treatment fad is additives based on Teflon—Du Pont's trade name for polytetrafluorethylene, or PTFE. Typical is Slick 50, which is PTFE powder in a carrier fluid of conventional or, optionally, of synthetic oil. Slick 50 is manufactured by Petrolon, which advertises heavily and is happy to provide expensive packages of PR materials for books like this. Petrolon claims great in-

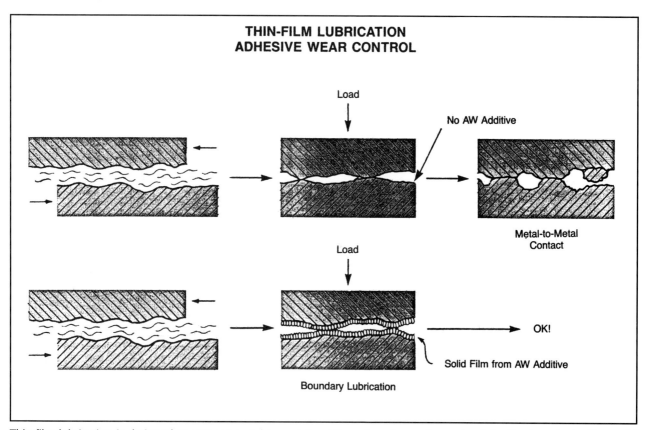

Thin-film lubrication is designed to prevent metal-to-metal welding of engine parts coming into direct contact with a solid film of organic coating. *Chevron*

creases in horsepower, reduced engine wear, quicker starting times, reduced engine operating temperatures, smoother running, and better fuel economy.

Independent tests by organizations, including the U.S. Department of Defense, indicate these types of claims have no credibility. What is worse, there appear to be documented cases of PTFE actually damaging engines. In one well-known case, PTFE additive had completely covered the engine's oil filter element and blocked the oil pump sump pickup, leading to oil starvation at the engine bearings, and significant wear damage. In addition, the piston tops had a liberal coating of PTFE flakes. There is also a documented case of an aircraft suffering engine failure in the mountains over Yellowstone National Park after a PTFE additive had completely blocked the oil pump sump. Oil analysis samples from other aircraft using PTFE additives showed a 10 to 20 percent increase in iron and aluminum in the oil. The theory about these catastrophes is that extreme cold weather causes the formation of wax crystals in the oil which caused the PTFE powder to congeal and plug the sump.

Forgetting for a moment about the possibility of a PTFE disaster, PTFE additives appear to lose 75 percent of the PTFE into the oil filter in fifteen minutes running. And although gears and engine parts which have received a special factory Teflon treatment can show reduced wear, the Teflon coating must be chemically bonded with special methods at temperatures above 800 degrees in order to adhere. "[Even] if it were possible to make PTFE stick to the internal parts," says one research engineer for a major oil company, "it would be readily scraped off by the constant motion of the piston rings against the cylinder walls..."

Du Pont quit selling Teflon powder for use as an oil additive around 1980, citing lack of any evidence that it was any use as an oil additive. Controlled studies by BMW of the effect of PTFE-resin additives found a significant increase in wear metals in the oil of the PTFE-treated engines, compared to a control group, and tear-down revealed accelerated wear of the cylinder walls, pistons, camshafts, and lifters.

The Army agrees: "The claims made for these [PTFE] products are supported by technically unsound data. When technically competent information is available, the claims cannot be satisfactorily substantiated. Based on these findings, it does not appear that significant benefits can be derived from the use of PTFE-containing engine oils or engine oil additives..." The Army report mentions that claims for

PTFE gear and grease products are also not supported by data. "Until such information is furnished," says the Army, "it is strongly recommended that PTFE-containing lubricants not be used in either the Military administrative or combat/tactical fleets."

Purging PTFE

The automotive newsletter *Nutz and Boltz* offers the following tips for getting PTFE sludge out of your engine:

1) To determine if there is internal damage to the engine, have the oil analyzed by a lab for oil metal content. If there is evidence of damage, it may be rebuild time.

2) Immediately add a quart of synthetic oil to counteract the possible wear caused by PTFE plugging oil passages. Synthetic oil is an effective detergent and will begin to loosen and break up the caked PTFE.

3) Change the oil and filter three times, every 1,000–2,000 miles, using 5W-30 oil with at least one quart being synthetic. Overfill the engine by 1/2 quart to help put the PTFE in suspension.

4) If your engine has less than five years or 60,000 miles on it, use an engine flush to help dissolve the PTFE resin buildup inside the engine. Using the flush on an older engine might harm an internal seal.

5) Use high-quality filters to help collect the PTFE particles from the oil. Consider installing a bypass filtration system (they cost $100 or so), to keep the oil analytically clean.

6) Between oil changes, get the engine thoroughly hot by running it at highway speeds for at least 30 minutes, which will help the detergent-dispersants in the motor oil to clean the engine. They require high temperatures to work.

Synthetic Oils

Unlike mineral oils, which have been distilled from the stew of various hydrocarbon molecules in crude oil, and still consist of dozens of varied hydrocarbons, synthetics are "designer" oils, chemically re-assembled or synthesized from homogeneous low-weight molecules into higher weight molecules tailored to a specific lubricative purpose. Synthetic oil production typically begins by producing ethylene from crude petroleum or natural gas. Ethylene is turned into alphaolefins (low-weight synthetic hydrocarbons) that are then polymerized into polyalphaolefins of heavier molecular weight. Synthetic base stocks also contain esters, which are essentially acid/alcohol compounds fully compatible with oil additives. They must be carefully blended with polyalphaolefins, which are thermally stable over a wide range but not particularly compatible with additives.

Why Synthetics are Better

Synthetics have none of the wax of which mineral oil, for economic reasons, cannot feasibly be 100 percent-free. Synthetic oils' pour point is -60 degrees F or below, while mineral oils only pour to about -40 F, but they are also formulated for improved high temperature stability. With their tightly coherent molecular structure, synthetics have inherently better resistance to thermal breakdown, and require less VI-improver (perhaps none) to maintain viscosity and integrity at high temperatures, instead of degrading into vapor and sludge. Synthetics not only maintain their thickness better at high temperatures, but they are also more resistant to longer term breakdown and thickening by repeated "cooking" under heavy loads.

All oils meet API standards with some margin, but synthetics tend not only to meet but to exceed the highest standards for passenger car motor oils. In the API test a 350 V-8 is tested under heavy load for 64 hours with 300 degree oil, and then disassembled and checked for internal wear and cleanliness. Synthetic oils' customized molecules are formulated with structures optimal for reduction of internal friction: Synthetics are actually slipperier than ordinary mineral oils.

Synthetics and Racing

The newest synthetics like Mobil 1 and other 100 percent synthetics from Quaker State, Valvoline, Texaco, and so forth have produced dramatic results in test engines, logging 200,000 miles with no significant wear. These oils are all excellent, although Mobil's testing indicates that there are still some "backyard chemists" operating in the synthetic oil business, and that good conventional oils are superior to some cheaper synthetics. The best synthetics are so good that virtually all race engine builders are now using synthetics. For most people, the main problem with synthetics is the price—anywhere from $2.99 to $9 a quart.

Lubrication Problems in Gasoline-Fueled Engines

Problems with gasoline engine lubrication have to do with the following:

1) Low-temperature or light-duty operation resulting in excessive contamination by partially burned fuel fragments and other blow-by products.

2) High-temperature lubrication oxidation, resulting in excessive engine rust and sludging that can lead, among other things, to oil thickening.

3) Valve train wear aggravated by increases in cam lift and valve spring loads necessary to achieve high volumetric efficiency and high engine speeds.

4) Extended oil change intervals.

5) Problems from emissions control devices like positive crankcase ventilation, thermactor, and exhaust gas recirculation.

6) Turbocharging.

In spite of the trend toward extended oil change intervals on original equipment new cars, oil and filters are still cheap compared to the cost of an engine rebuild. Enthusiasts should change gasoline engine oil frequently (every 3,000–5,000 miles) and should use the best synthetic oil available in cars, beginning at least 2,000 miles after the beginning of a break-in period. They should not use racing motor oils, which may contain additive packages which will not protect a gasoline engine for street operation.

Racing Gasoline Lubricants

There are a number of crankcase oils designed specifically for use in gasoline-fueled race cars. Oils designed for racing use do not have to stand up to long intervals between oil changes, and the additive package similarly does not have to fight the same kind of rust, gum, sludge, and neglect that street crankcase oils must withstand. They can be optimized for preventing wear. Sunoco Custom Lubricants' Dynatech 5W-40 and 20W-50 synthetic engine oils are designed to combat excessive wear during cold startup and under the extreme heat of racing conditions. Sunoco claims that Dynatech's lubricity characteristics and friction modifier additives may reduce internal engine friction, resulting in horsepower gains. Sunoco claims that 5W-40 oil is compatible with aluminum cylinder cases and high silicon, low expansion pistons, with clearances of only 0.001 inches. Sunoco recommends Dynatech 20W-50 synthetic with Sunoco racing fuel.

Unocal High Performance Engine Oil (a mineral oil) is used in NASCAR racing, and is designed for severe and/or high temperature conditions, although Unocal claims it works well in stop-and-go driving. The three Unocal high performance oils (10W-30, 10W-40, and 20W-50) are turbo-tested SH oils that meet new car warranty specs. The oils are friction-modified for more horsepower and economy. They include an additive package designed to minimize corrosion, rust, sludge, and varnish formation.

Lubricants for Alcohol-Fueled Engines

Sunoco Custom Lubricants and others provide crankcase oils specifically designed for alcohol-fueled vehicles. Alky 20W-50 engine oil is designed to fight problems associated with alcohol fuels such as increased camshaft wear and oil dilution, which can substantially reduce engine life. The corrosive nature of alcohol, particularly methanol, and harmful by-products of combustion require special oil. Alky 20W-50 is designed for engines fueled with 100 percent alcohol, and includes dispersants to protect the engine from harmful acids that tend to form in the crankcase. Sunoco recommends using Alky with their SunTop alcohol fuel additive, which includes detergents, dispersants, and lubricity agents. SunTop is designed to be mixed in a ratio of 1 quart to a 55-gallon drum, 3 ounces per 5-gallon pail, or 0.5 ounces per gallon.

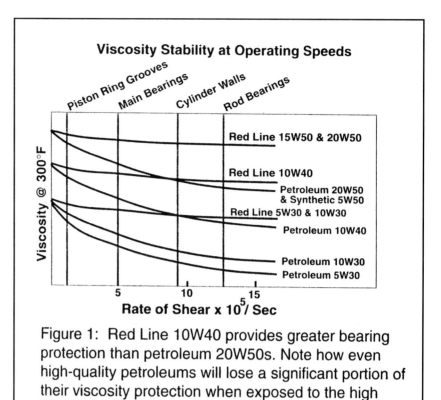

Figure 1: Red Line 10W40 provides greater bearing protection than petroleum 20W50s. Note how even high-quality petroleums will lose a significant portion of their viscosity protection when exposed to the high shear forces in bearings and on cylinder walls.

Stabilty of oil at high shear rates is vital. Synthetic lubricants like those from Red Line provide greater bearing protection than higher viscosity mineral oils. *Chevron*

Lubricants for Nitro-Fueled Engines

Top-Fuel Dragster and Top Fuel Funny Cars burning nearly pure nitromethane produce extreme cylinder pressures and bearing loads beyond the normal design limits of the engines with conventional oils. As Sunoco Custom Lubricants points out, Top Fuel engines' bearing surface areas originally designed to carry loads of several hundred horsepower may carry loads of several thousand horsepower. In addition, Top Fuel motors running extremely rich air-fuel mixtures to fight detonation also experience very high levels of oil dilution. Oils like Sunoco Nitro 60 include extremely high levels of dispersants to withstand high levels of fuel dilution. They are manufactured to SAE 60 viscosity for good thick-film lubrication under demanding conditions, and also contain high levels of antiwear (thin-film) additives and extreme pressure additives to "help protect the engine in extremely hostile applications." Oils like Nitro 60 could be duplicated by a user blending additives with high-viscosity oils, but original equipment oils like Nitro 60 take away the guesswork.

Lubricants for Propane and NG-Fueled Engines

Propane and natural gas engines do not have to contend with sulphur or gum deposits from gasoline, so the barium and calcium oil additives used in high detergent multiviscosity oils have nothing to combine with, and can themselves burn and cause valve and spark plug deposits. High thermal loading, possibly aggravated by incorrect mixtures or spark advance curves can worsen oil additive burning. Special low-additive oils are available for propane-fueled engines. Some propane and natural gas equipment manufacturers recommend against using ashless dispersant oils (including synthetics, which are classified as ashless) on propane vehicles because ashless oils can contribute to valve recession on older engines without hardened valve seats due to the reduced carbon content of propane. Propane and natural gas vehicles can often do with less frequent oil and filter changes.

Diesel Engine Lubrication

The trend in diesel engines has been to increase the output per unit engine weight, mainly by turbocharging or increasing the operational speed. This has resulted in a need for improved lubricants. Test severity levels have been increased: higher ring-belt area temperatures and higher power output have increased the stress on anti-oxidation and thermal stability of the lubricant. As diesel power has increased, diesel engine manufacturers have needed improved ring and bearing wear control and reduced oil consumption. The need to control emissions and improve fuel economy has caused some diesel manufacturers to consider EGR. An apparent increase in diesel fuel sulfur levels has made the lubrication task more difficult, as has the manufacturers' desire to increase oil drain intervals. Given the superior efficiency of diesel engines (40 percent versus 25 percent in spark ignition engines), many more cars are now powered by diesel engines. Studies have shown that the increased valve wear in diesel engines seems to be due to diesel soot or carbon contamination of the crankcase oil, in which the antiwear agent is absorbed by soot, reducing protective antiwear surface coating. The above factors make it incumbent for diesel users to use modern top-quality diesel crankcase lubricants.

Oil Analysis

There are laboratories with the capability to analyze engine oil in order to deduce the internal condition of an engine without disassembling it. This procedure is routine for aircraft, but is becoming much more common for construction fleets, and can be useful for automotive enthusiasts and racers. Labs analyze the oil for metal content, and based on the amount of bearing metal, ring metal, and so forth, can make educated guesses about the condition of the engine. This is most useful for a known type of engine, where normal wear has been studied. It is very important to follow the directions of the lab in terms of length of interval since last oil change, and so on.

4

Engines and Fuels

Fuels engineer Tom Hart points out that fuel development tends to follow developments in engine technology, rather than the reverse. Someone comes up with a great new design that requires special fuel; chemists and engineers then concoct fuels optimized for the new powerplant. How does engine design affect fuel requirements?

Normal combustion in a spark ignition engine occurs after the spark plug ignites the compressed air-fuel mixture, starting a flame front which spreads out in a wave from the plug in all directions, something like dry grass burning in a field. It takes time for the flame to spread in the air-fuel mixture, exactly as it takes time for a field of dry grass to burn away from where the fire was started. During normal combustion, the flame front moves smoothly and evenly across the combustion chamber until it reaches the other side.

Special cameras recording the combustion event through a porthole in a cylinder reveal that flame-front speeds for gasoline-air mixtures vary from 20 feet per second to over 150 feet per second, depending on air-fuel ratio, density, compression ratio, turbulence among the charge gases, and combustion chamber design. Flame speed is much slower in a rich or a lean mixture. The slower the flame speed, the greater the chance of abnormal combustion.

High-compression engines make more power because they are more efficient at getting energy from the air-fuel charge mixture. Normally, combustion chamber pressure rise is 3.5 to 4 times the initial compression pressure. The difference in pressure gain in an 8.5:1 compression engine versus a 10:1 compression engine can easily be 50 percent—as much as 250 psi. A typical 160 psi compression pressure rises quickly to over 600 psi following combustion. Pressure rise rates can exceed 20 psi per degree of crankshaft rotation. Cylinder pressures are highest at wide open throttle, lowest at idle and light cruise when intake manifold vacuum is very high and engine VE very low. Combustion proceeds faster under higher compression.

Turbulence is carefully designed into combustion chambers. Some swirl will produce greater flame speed and more efficient

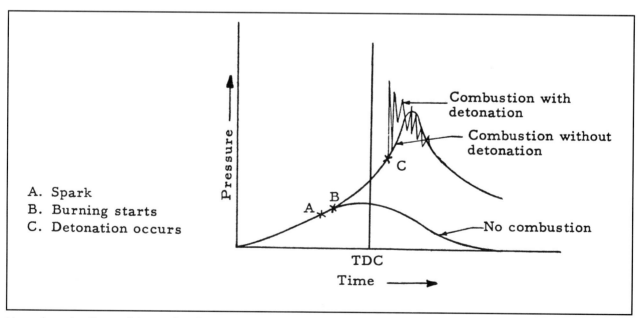

Normal combustion and detonation. Detonation produces severe pressure shock waves at a very high rate which can cause high rates of wear or even catastrophic failure. Detonation does not itself cause power loss, but greatly increases the likelihood of pre-ignition, which does cause large power drops. *Automotive Engineering*

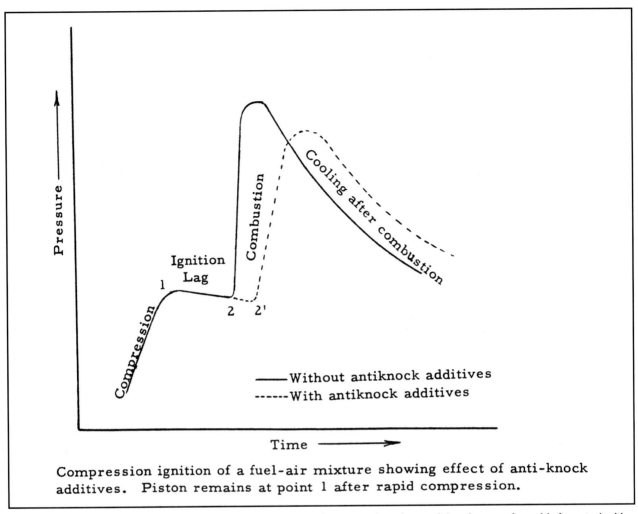

Compression ignition of a fuel-air mixture showing effect of anti-knock additives. Piston remains at point 1 after rapid compression.

Antiknock additives help prevent detonation by delaying the time between when the conditions become favorable for auto-ignition and its actual onset—allowing the time for normal combustion to finish before detonation can get started. *Automotive Engineering*

combustion, and even help resist knock by spreading the flame front more quickly. Too much swirl will produce undesirable disturbance of the quench layer around the combustion chamber and increase heat loss into the cylinder walls.

The actual performance of fuel in an engine is closely related to the release of energy during the combustion process. This is affected by the reactants and products of combustion dissociating (molecules splitting into smaller molecules or single atoms or ions), and by heat transfer through the combustion chamber walls. At wide-open throttle at high speed, heat transfer amounts to about 15 percent of the fuel's thermal energy. This heat loss decreases with increased brake mean effective pressure in the cylinder. The heat transfer is maximized at stoichiometric air-fuel ratios and decreases with rich mixtures. Heat transfer also depends on gas turbulence near the combustion chamber surfaces and increases with knock and other violent combustion conditions.

Given that engines induct a fairly constant mass of air for a given speed, manifold pressure, camshaft timing, and intake temperature, the need to improve spark-ignition engine power for racing has led to two main strategies in an effort to increase the amount of fuel that can be burned in the combustion chamber. The first has been to increase oxygen intake by supercharging, or by supplementary oxygen injection—usually in the form of nitrous oxide. The second has been to search for fuels with high specific energy.

In addition to the octane number requirement of an engine in order to avoid detonation, high-performance engines have other performance requirements for fuels, including volatility, flame speed, weight, and volume.

Octane Number Requirement (ONR)

When an engine knocks or detonates, combustion begins normally with the flame front burning smoothly through the air-fuel mixture, but as pressure and temperatures rise as combustion proceeds, at a certain point the remaining "end" gases explode violently all at once, rather than burning evenly. The resulting high-pressure shock waves in the combustion chamber can accelerate wear or even cause catastrophic failure.

Pre-ignition is another form of abnormal combustion in which the air-fuel mixture is ignited by something other than the spark plug, such as glowing combustion chamber deposits, sharp edges or burrs on the head or block, or an overheated spark plug electrode. Heavy, prolonged knock can generate hot spots that cause surface ignition, which is the most damaging side effect of knock. Surface ignition which occurs prior to the plug firing is called pre-ignition, and surface ignition occuring after the plug fires is called post-ignition. Pre-ignition causes ignition timing to be advanced by some inexact amount, so the upward movement of the piston on the compression stroke is opposed by the too-early high combustion pressures, resulting in power loss, engine roughness, and severe heating of the piston crown. Pre-ignition can lead to knock, or vice versa. Dieseling, or "run on," is usually caused by compression ignition of the air-fuel mixture by high temperatures, but can be caused by surface ignition. A fuel's octane rating represents its ability to resist detonation. The possibility that detonation may lead to pre-ignition also means that the octane rating of a fuel—especially gasoline, or a fuel like gasoline—is loosely related to its ability to resist pre-ignition.

Apparently identical vehicles coming off the same assembly line can have octane requirements that vary by as much as ten octane numbers. The octane number requirement (ONR) for the engine in a particular vehicle is usually established by making a series of wide-open throttle accelerations at standard spark timing using primary reference fuels with successively lower octane ratings, until a fuel is found that produces trace knock.

Factors influencing an engine's octane requirements are effective compression ratio, atmospheric pressure, absolute humidity, air temperature, air-fuel ratio,

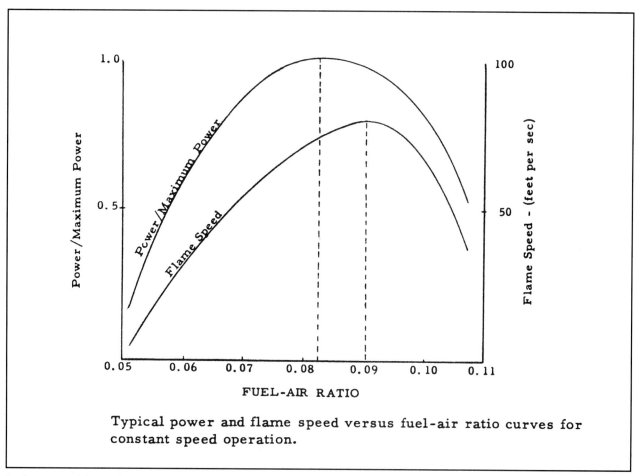

Typical power and flame speed versus fuel-air ratio curves for constant speed operation.

Peak power with gasoline fuel occurs just rich of rich best torque, at an air-fuel ratio of just above 11:1. The cooling heat of vaporization of such rich mixtures along with the high flame speed help fight detonation. *Automotive Engineering*

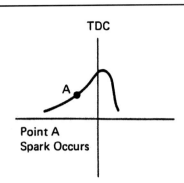

Normal Cylinder Pressure Graph.

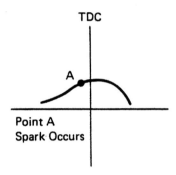

Combustion Pressure Graph Resulting from Late Ignition Timing.

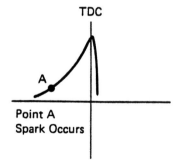

Combustion Pressure Graph Resulting from Early Ignition Timing.

Maximum power occurs when peak cylinder pressure occurs at 15 degrees ATDC. Late ignition timing produces inefficient combustion of a less dense mixture, while early timing produces high pressures as the piston is still moving upward on the compression stroke, falling off rapidly following top dead center.

variations in mixture distribution among the engine's cylinders, oil characteristics, spark timing, timing variations between individual cylinders, intake manifold temperature, coolant temperature, coolant characteristics, and combustion chamber hot spots.

The single most important engine characteristic demanding specific fuel characteristics is compression ratio, which increases the ONR +3 to +5 per one ratio increase, in the 8–11:1 CR range. High compression ratios squash the inlet air/fuel mixture into a more compact, dense mass, resulting in a faster burn rate, more heating (yet less heat loss into the combustion chamber surfaces), and consequent higher cylinder pressure. Turbochargers and superchargers produce effective compression ratios far above the nominal compression ratio by pumping additional mixture into the cylinder under pressure. The result is a more dense charge that burns faster and produces more pressure against the piston and an increased tendency to knock.

High peak cylinder pressures and temperatures resulting from high compression can also produce more NO_x pollutants. Lower compression ratios raise the fuel requirements at idle because there is more clearance volume in the combustion chamber which dilutes the intake charge. And because fuel is still burning longer as the piston descends, lower CRs raise the exhaust temperature and increase stress on the cooling system.

Until 1970, high-performance cars often had compression ratios of up to 11 or 12 to 1. This was easily handled with the then readily available gasolines with octane numbers in the 98–99 ((R+M)/2) range. By 1972, engines were running compression ratios with 8–8.5:1. In the 1980s and 1990s, compression ratios in computer-controlled fuel injected vehicles were again up in the 9.0–11:1 area, due to fuel injection's ability to support higher compression ratios without detonation through precise air-fuel ratio control. Race car engines typically run even higher compression ratios. In air-unlimited engines, maximum compression ratios with gasoline run in the 14–17:1 range.

Ratios above 14:1 demand not only extremely high octane fuel (which might or might not be gasoline), but also uniform coolant temperature around all cylinders, low coolant temperature, uniform fuel distribution to all cylinders, retarded timing under maximum power, very rich mixtures, and probably individual cylinder optimization of spark timing and air-fuel ratio.

Air-Fuel Ratios

Air-fuel ratio has a major impact on engine ONR, increasing octane requirements by +2 per one increase in ratio (say from 8:1 to 9:1). Ideally, air-fuel ratio should vary not only according to loading but according to the amount of air present in a particular cylinder at a particular time. Richer air-fuel ratios combat knock by the cooling effect of the heat of vaporization of liquid fuels, and via a set of related factors. The volatility of fuels affects not only octane number requirement but drivability in general.

The chemically ideal, or stoichiometric, air-gasoline mixture, at which all air and fuel are consumed in combustion, occurs with about 14.6 parts air and 1 part fuel, by weight. Stoichiometric mixtures vary according to fuel, from a low of nitromethane at, say, 1.7:1, to methanol's 6.45:1, ethanol's 9:1, up to gasoline at, typically, 14.6:1, and beyond; natural gas and propane are in the range of 15.5–16.5:1. Mixtures with a greater percentage of air than stoichiometric are called lean mixtures; mixtures with an excess of fuel are said to be rich.

At high loading and wide-open throttle, richer mixtures give better power by making sure that all air molecules in the combustion chamber have fuel available to burn. At wide open throttle, where the objective is maximum power, all four-cycle gasoline engines require mixtures that fall between lean and rich best torque, in the 11.5 to 13.3 range, in the case of gasoline. Since this best torque mixture spread narrows at higher speeds, a good goal for naturally aspirated engines is 12 to 12.5 to 1, and perhaps richer if excess fuel is being used for cooling in a supercharged engine.

Typical gasoline-air mixtures giving best drivability are in the range of 13.0 to 14.5 to 1, depending on speed and loading. At higher engine speeds, reverse pulsing through a carb in engines with racing cams tends to richen the mixture as reversion gases pass through the venturi twice. Naturally, this is not a problem with fuel injection.

EFI engines are not susceptible to hot weather percolation in which fuel boils in the carb, flooding into the manifold. This tendency is somewhat dependent on the distillation curve of the fuel, which varies according to the weather for which it is intended. Hot weather percolation can be aggravated by high vapor pressures. Similarly, vapor lock, produced when the fuel boils in the line before the carburetor, eventually uncovering the main jets as fuel level drops, is not a problem with EFI.

A major advantage of computer-controlled engines is that variations in speed, load, and other parameters translate into spark advance and fuel-air ratio values that can be completely independent of those values under other operating conditions (and could vary quite abruptly, if necessary), something that is impossible with older mechanical control systems. Multiport injection—as distinct from throttle body injection—has the further advantage of eliminating the problems of inconsistent distribution and wet mixtures in the intake manifold that are associated with less than one carb per cylinder. This often results in improved cold running, improved throttle response under all conditions, and improved fuel economy without drivability problems.

Computer engine management with individual cylinder adjustable electronic port fuel injection really shines on extremely high output engines with high effective compression ratios. Dallas supercar builder Bob Norwood points out that for peak power, any engine must be treated as a gang of single cylinder engines, each of which should be optimized for peak power.

On NASCAR engines, which are restricted to single-point carburetion, this is done by custom-modifying the compression ratio, rocker arms, and cam lobes for individual cylinders. But using sequential programmable EFI with fuel and ignition calibration on an individual cylinder basis, optimal timing and mixture can be provided, whatever the volumetric efficiency of each "single cylinder engine."

Individual cylinder calibration is good for 50 horsepower on a high output small-block Chevy, says Norwood. Not only does it improve power, it also saves engines from knock damage while allowing a higher state of tune. By contrast, common practice on street engines with carburetors or batch fuel injection is to tune to the leanest cylinder. A Chevy V-8 with its dog-leg ports typically varies 10 percent in airflow between best and worst cylinders, and it is typical that V-type motors respond better than in-line motors to individual cylinder calibration. On a dyno, best and worst cylinders within 100 degrees exhaust gas temperature (EGT) of each other is considered excellent. Leaning out an engine on a dyno or track will frequently make more power until suddenly one or two cylinders die. "You always burn one or two cylinders," says Norwood. "In the meantime, the other slackers are running rich." The latest generation of electronic engine management systems like Enjectec's EGT-feedback EFI equalizes EGT with individual cylinder calibration, simultaneously providing best power everywhere. Each cylinder can reach maximum power at very high compression ratio with the best high-octane gasoline.

Testing by Smokey Yunick for *Circle Track* showed that the small-block Chevy test engine was prone to detonation on three of the eight cylinders. This was thought to be due to unequal distribution or coolant circulation which caused those cylinders to run hotter. "Now we could have cut these three cylinders to 14:1," writes Smokey, "and maybe run the other five cylinders at 15:1 and gained some power. Don't know, but I do know this, you are gonna get a safer, more powerful engine if you adjust the compression, timing, and distribution to each cylinder."

Spark Advance, Flame Speed, VE, and Air-Fuel Ratio

Spark advance, which is optimally timed to achieve best torque by producing peak cylinder pressure at about 15 degrees ATDC, increases octane requirements by 1/2 to 3/4 of an octane number per degree of advance. Spark advance increases cylinder pressure and allows more time for detonation to occur.

Engine speed and fuel burn characteristics affect ignition timing requirements. As an engine turns faster, the spark plug must fire sooner in order to allow time for the air-fuel mixture to ignite and achieve a high burn rate and maximum cylinder pressure by the time the piston is positioned to produce best torque. The amount of additional advance depends not only on engine speed but also on fuel-air flame speed, which in turn depends not only on the type of fuel but also on operating conditions which change dynamically, such as air-fuel mixture. Changes in engine load, for example, mean differing throttle positions, which affects VE, and so the optimal air-fuel mixture. That, in turn, requires modified ignition timing.

An important factor which affects VE, and potentially flame speed, is valve timing. Remember, a denser mixture burns more quickly, and a leaner mixture requires more time to burn. Valve timing has a great effect on the speeds at which an engine develops its best power and torque. More lift and overlap allows the engine to breathe more efficiently at high speeds; however, the engine may be hard to start, idle badly, bog on off-idle acceleration, and produce bad low-speed torque. This occurs for several reasons. Increased valve overlap allows some exhaust gases still in the cylinder at higher than

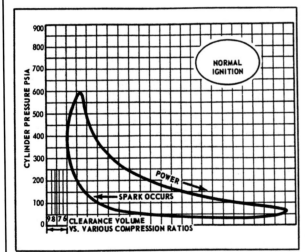

Theoretical pressure-volume diagram for a four-stroke cycle engine: The tip at lower right represents the intake stroke. Following the lower curve leftward, we see that compression brings a temperature rise. The pressure is increased as cylinder volume is diminished and skyrockets when the mixture is ignited. Peak temperature is reached before the point of maximum pressure. Then the pressure and temperature decline during the power stroke until the exhaust valve opens. Engine design compromises resulting in heat and friction losses keep the curve theoretical—but fuel injection engines come closer than carbureted engines.

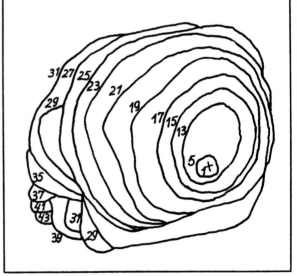

Flame front travel: The flame front spreads from the spark plug (x). By filming the combustion process through a "window" and using a mirror, it has been possible to map the progress of the flame front. This is from a Mercedes-Benz engine with parallel, canted overhead valves and fuel injection. The numbers indicate the sequence in milliseconds, revealing a critical slowing-down in the quench area adjacent to the inlet valve.

Pressure-volume and flame front mapping.

atmospheric pressure to rush into the intake manifold, diluting the inlet charge. Gross exhaust gas dilution of the air-fuel mixture at idle requires a lot of spark advance and a mixture as rich as 11.5 to 1 to counteract the lumpy uneven idle resulting from partial burning and misfires on some cycles. Valve overlap also hurts idle and low-speed performance by lowering manifold vacuum. Since lower atmospheric pressure of high vacuum tends to keep fuel vaporized better, racing cams with low vacuum may have distribution problems and a wandering air-fuel mixture at idle. This is not a problem with fuel injection.

With medium speeds and loading, the bad effects of big cams diminish, resulting in less charge dilution, allowing the engine to happily burn air-gasoline mixtures of 14 to 15:1 and higher. At the leaner end, additional spark advance is required to counteract slow burning of lean mixtures. Hot cams may produce problems for carburetted vehicles when changed engine vacuum causes the power valve to open at the wrong time. Changed vacuum could affect speed-density fuel-injection systems but would have no effect on Mass Air Flow (MAF)-sensed EFI. Where air-fuel mixture are inconsistent or poorly atomized, flammability suffers, affecting many other variables.

With a big cam, the spark advance at full throttle can be aggressive and quick. Low VE at low rpm results in slow combustion and exhaust dilution lowering combustion temperatures and the tendency to knock. Part throttle advance on big cam engines can also be aggressive due to these same flame speed reductions resulting from exhaust dilution of the inlet charge due to valve overlap.

Turbulence and swirl are extremely important factors in flame speed—more important, within limits, than mixture strength or exact fuel composition. Automotive engineers have long made use of induction systems and combustion chamber geometry to induce swirl

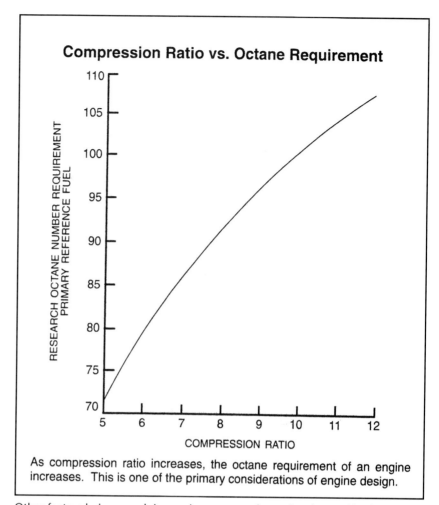

As compression ratio increases, the octane requirement of an engine increases. This is one of the primary considerations of engine design.

Other factors being equal, increasing compression ratios demand higher octane fuels, since pro-knock heat and temperature factors from high compression override higher flame speed benefits.

or turbulence to enhance flame speed and, consequently, antiknock characteristics of an engine. Wedge head engines, with a large quench area, have long been known to induce turbulence or swirl as intake gases are forced out of the quench area as the piston approaches top dead center.

In the 1970s, automotive engineers began to de-tune engines to meet increasingly tough standards. They began to retard the ignition timing at idle, for example, sometimes locking out vacuum advance in lower gears or at normal operating temperature, allowing more advance if the engine was cold or overheating. Since oxides of nitrogen are formed when free nitrogen combines with oxygen at high temperature and pressure, retarded spark reduces NO_x emissions by lowering peak combustion temperature and pressure. This strategy also reduces hydrocarbon emissions.

However, retarded spark combustion is less efficient, causing poorer fuel economy and higher operating temperatures, as heat energy escapes through the cylinder walls into the coolant. The cooling system is stressed as it struggles to remove the greater waste heat during retarded spark conditions. Fuel economy is also hurt since some of the fuel is still burning as it blows out the exhaust valve, necessitating richer idle and main jetting to get decent off-idle performance, and because if the mixture becomes too lean, higher combustion temperatures will defeat the purpose of ignition retard, producing more NO_x. Inefficiency of combustion under these conditions also requires the throttle to be held open further for a reasonable idle speed, which, combined with the higher operating temperatures, can lead to dieseling. (This is not a problem in fuel-injected engines which immediately cut off fuel flow when the key is switched off.) By removing pollutants from exhaust gas, three-way catalysts tend to allow more ignition advance at idle and part throttle.

Similarly, gasoline engines converted to run on propane, natural gas, or alcohol require a different timing curve due to variations in combustion flame speed of the air-fuel mixture (see chapters on non-gasoline fuels).

Jacket Coolant Temperature

Increases in coolant temperature increase octane requirements by roughly one octane number per 10 degree increase from 160 to 180 degrees F.

Ambient Temperature

Inlet air temperature increases octane requirements by .5 octane number per 10 degree increase. Temperature affects fuel performance in several ways. Colder air is more dense than hotter air, affecting cylinder pressure. Colder air inhibits fuel vaporization. Hotter air directly raises combustion temperatures, which increases the possibility of knock.

Engines will make noticeably more power on a cold day because the cold dense air increases engine volumetric efficiency, filling the cylinders with more molecules of air. As you'd expect, racing automotive engine designs always endeavor to keep inlet air as cold as possible, and even stock street cars often make use of cold air inlets, since each 11 degree F increase reduces air density one percent.

This is bad news for a carb, which has no way of compensating, other than by manual jet changes (typically one size per 40 degree temperature change).

Cold engines require enrichment because only the lightest fractions of liquid fuel may vaporize at colder temperatures, while the rest exist as globules or drops of fuel that are not mixed well with air. At temperatures below zero, air-gasoline mixture may optimally be as low as 4 to 1 for best drivability, and during cranking as rich as 1.5 to 1!

In a cold engine, most of a liquid fuel will be wasted, and most air pollution is produced by cold vehicles.

The actual vaporization/distillation curve of various gasolines and fuels differs, depending on the purpose for which the fuel is designed, and the oil companies change their gasoline formulation to increase the vapor pressure in cold weather—which can lead to a rash of vapor lock in sudden warm spells in winter. Gaseous fuels like propane and methane do not require cold running enrichment; on the other hand, they have no evaporative cooling effect once in gaseous form.

Absolute Humidity

Humidity, which increases the amount of water vapor in the combustion chamber, decreases octane requirement by 1/3 octane number per 10 grains increase per pound of air.

Altitude, Air Density, Fueling, and Turbocharging

Increasing altitude reduces octane number requirements by about 1.5 octane numbers per 1,000 feet above sea level. Supercharging, which provides a denser breathing atmosphere, has the opposite effect, and further increases ONR due to the tendency of superchargers to heat inlet air as well as increasing its density.

Air density varies with temperature, altitude, and weather conditions. For a given pressure, hot air is less dense, as is air at higher elevations or with a higher relative humidity. Intake system layout can have a great effect on the volumetric efficiency of the engine by affecting the density of the air the engine is breathing. Air cleaners that suck in hot engine compartment air will reduce the engine's output and should be modified to breathe fresh cold air from outside for maximum performance. Intake manifolds which heat the air will produce less dense air, although a properly designed heated intake manifold will very quickly be cooled by intake air at high speed and will probably improve distribution at part throttle and idle.

Turbochargers, which output heated, turbulent air, are known to enhance good air-fuel mixing and to enhance atomization.

Combustion Chamber Deposits

Depending on engine design, fuel, lubricant, and operating conditions, combustion chamber deposits can increase ON by 1 to 13 numbers.

Gasoline

5

What we refer to as gasoline is not generally a single substance—not one pure specific hydrocarbon. Gasoline is a blend of various olefinic, paraffinic, naphthenic, and aromatic hydrocarbons (plus additives), formed during the refining of crude oil, or synthetically manufactured. The octane and other characteristics of a particular gasoline depend on the composition of the blend.

Gasoline's heat value can be directly converted to useful work. One gallon of typical gasoline contains roughly 115,000 BTUs of energy (in metric units, 42.7 MJ/kg), and is capable of developing 89,000,000 foot/pounds of work. This amounts to 2,700 horsepower for one minute or 45 horsepower for one hour. Since modern reciprocating engines are only about 27 percent efficient, a gallon of gasoline, practically speaking, can produce about 13 horsepower for an hour.

Gasoline has the highest heating value of the common high-performance motor fuels, with over twice the energy per pound of alcohol, for example—nearly four times that of nitromethane. When weapons researchers chose a base fuel for fuel-air explosives, they chose gasoline. When there's plenty of oxygen available, it's hard to beat gasoline for making energy.

History

Gasoline was once considered a useless by-product of refining crude oil, and was run off in ditches as waste. It is uncertain who discovered it, but Joshua Merrill, working in Boston in 1863, may have isolated gasoline while attempting to refine kerosene. Gasoline was used commercially early on in air-gas machines to produce fuel that could be piped and burned as a source of light in street lamps. In 1876, it was the fuel used in the original four-stroke internal combustion engine built by Nikolaus Otto, in Germany.

Gasoline came into its own in the early twentieth century as a motor fuel. In the years between 1900 and 1920, gasoline ceased to be a by-product of refining other hydro-

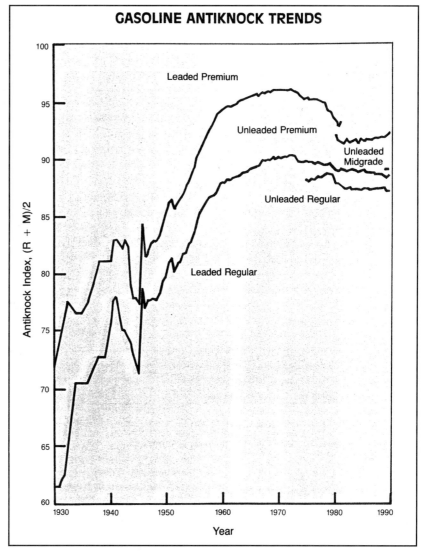

Historical octane trends. Leaded premium disappeared in the early 1980s in the U.S., while leaded regular was completely phased out in 1995 for environmental reasons. The first widespread use of unleaded fuels occurred in the mid-1970s. *Chevron*

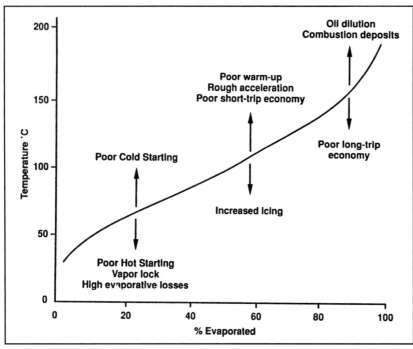

Gasolines with higher volatility (below the curve) provide easier starting and warmup operation but increased tendency to vapor lock and evaporative losses in hot weather. Fuels with lower volatility (above the curve) provide lower evaporative losses and help prevent vapor lock, plus fuel economy is better for short trips. However, low volatility gasolines provide harder starting and can increase deposits and oil dilution during warmup plus increased emission in some cases. *Chevron*

carbons and was intentionally refined as a motor fuel. In 1913, a process called thermal cracking was invented to increase gasoline yield and quality by converting useless heavy crude oil fractions into lighter materials that boiled in the gasoline range. The thermal cracking process used heat and pressure to break down and rearrange heavy molecules into the light, volatile hydrocarbon components of gasoline.

Early engines were large and slow, and had low compression ratios. Gasoline blends of extremely low octane used in such engines in farm equipment were referred to by the generic term "tractor fuel." Gasolines had little resistance to detonation, but needed little in these early machines. However, as engines increased in speed and performance, higher performance gasoline with higher octane ratings became increasingly important.

Following WWI, benzol and higher aromatics were blended

Gasoline Specifications and Their Importance

Specification	Importance
Antiknock Index (AKI)	
Research Octane Number (RON)	Low to medium speed knock and run-on
Motor Octane Number (MON)	High speed knock/Part-throttle knock
Fuel Volatility	
Vapor Liquid (V/L) Ratio	Vapor lock
Distillation	Cool weather driveability, hot start and hot driveability, vapor lock, evaporative losses, crankcase deposits, combustion chamber and spark plug deposits
Vapor Pressure (RVP)	Low temperature starting, evaporative losses, vapor lock
Copper Corrosivity	Fuel system corrosion
Stability	
Existent Gum	Induction system deposits, filter clogging
Oxidation Stability	Storage life
Sulfur Content	Exhaust emissions, engine deposits and engine wear
Metallic Additives (lead and others)	Catalyst deterioration (unleaded vehicles)
Temperature for Phase Separation	Water tolerance of blended fuels

with hydrocarbons like cyclohexane to improve octane levels. And gasoline additives became important. Discovered as an antiknock additive on December 9, 1921, by Thomas Midgley and Thomas Boyd of General Motors Research Corporation, tetraethyl lead (TEL) was found to have the ability to boost octane ratings considerably, although with diminishing returns. This has an interesting implication. Blending high-octane unleaded automotive fuel, especially fuel containing high levels of alkylate or isomerate, with leaded avgas can produce a result with higher octane rating than either component. In any case, in commercial gasoline lead usage grew until WWII, when military needs had priority, and again increased in the 1950s as compression ratios rose rapidly as part of efforts to improve power and economy. Lead use per gallon began to drop off in the late 1950s as improved refining methods produced better quality base components of gasoline. TEL use peaked around 1970.

There have been continuous efforts this century to produce better base stocks of gasoline. In the 1930s, thermal reforming was invented and used to increase octane quality of poor quality, low-octane gasoline components by cracking and reforming them into light, high-octane hydrocarbons. Coking, developed in this same time period, is an extreme version of thermal cracking in which poor quality, heavy petroleum fractions are made into gasoline through pyrolysis. Catalytic cracking was invented in 1936, using a catalyst at reduced temperatures compared to thermal cracking with pressure, to reduce the size of hydrocarbon molecules to gasoline size. Catalytic reforming, which converts low-octane hydrocarbons into higher-octane aromatics, arrived in 1940. Polymerization and alkylation (1937 and 1938) became available to combine small gaseous hydrocarbons into larger liquid high-octane gasoline-sized hydrocarbons.

WWII was a watershed period for high-performance piston engine fuels research. Both the Allies and the Axis powers were under enormous pressure to develop high-performance fighter and bomber aircraft in efforts to achieve air-superiority. In particular, high output, supercharged, air-cooled airplane engines required extremely high octane in order to meet performance and reliability standards. As Germany's petroleum fuels infrastructure was increasingly damaged by allied bombing, and access to eastern European oil fields was cut off by Russian advances in the latter days of the war, German chemists developed methods of producing synthetic gasolines and lubricants from coal, gases, and other hydrocarbons.

This was a unique period for gasoline development. Fuels research data was shared freely among automotive and petroleum companies doing war research in the U.S. Following the war, high-performance aircraft began using jet engines, and heavy trucks and armored vehicles were increasingly powered by diesel engines (not only for the torque/lugging characteristics of diesel engines, but because less flammable diesel fuel was less likely to incinerate the crew under combat conditions). But during the war, gasoline and piston engines had top priority.

During the war, in 1943, isomerization was developed to im-

Gasoline Additives

Additive	Purpose
Detergents/deposit control additives*	Eliminate or remove fuel system deposits
Anti-icers	Prevent fuel-line freeze up
Fluidizer oils	Used with deposit control additives to control intake valve deposits
Corrosion inhibitors	To minimize fuel system corrosion
Anti-oxidants	To minimize gum formation of stored gasoline
Metal deactivators	To minimize the effect of metal-based components that may occur in gasoline
Lead replacement additives	To minimize exhaust valve seat recession

* Deposit control additives can also control/reduce intake valve deposits

Focus of Fuel Reformulation

Potential Reformulation	Environmental Benefits
Reduction of sulfur	Reduces acid rain / improves vehicle catalyst efficiency
Reduction of benzene	Reduces incidence of cancer / reduces toxicity of emissions
Reduction of aromatics	Reduces photochemical reactivity/ smog formation
Reduction of olefins	Reduces photochemical reactivity/ smog formation
Reduction in fuel volatility	Reduces evaporative emissions/ smog formation
Increase oxygen content	Reduces tailpipe CO emissions / replaces toxic components
Alteration of distillation characteristics	Control of evaporative & tailpipe emissions
Deposit control additives	Reduce engine deposits thereby reducing exhaust emissions

prove octane by forming branched-chain isomers from normal straight-line paraffinic hydrocarbons of identical chemical composition. Much later, hydrocracking, developed in 1959, performed the same function as catalytic cracking, but more efficiently, in a hydrogen atmosphere.

After the war, the focus of high-performance gasoline development shifted away from avgas. Motor racing in the 1950s onward, and the street horsepower wars of the 1950s and 1960s, provided the incentives for high-performance fuels. Automotive gasoline octane ratings peaked in the late 1960s, when super premium fuels like Sunoco 260 and Chevron "White Pump" were available at the pump with Research octane ratings of 103+, for fueling stock high-performance engines with compression ratios over 11:1, such as certain big-block Corvettes and 427 Fords. These super premium fuels were highly leaded, with typically 3.5 grams of lead per gallon. The motor octane was 94–95, and (R+M)/2 was 98–99. Sunoco pumps sucked varying amounts of high-octane and low-octane fuel simultaneously from two separate tanks in order to create seven custom octane rating gasolines from "190" to "260."

Until the 1970s, refiners altered the composition of gasoline in response to technological advances in refining and engine technology, and changes in end-user demand. More recent changes in gasoline composition have been driven by environmental considerations. Unleaded gasoline was introduced widely in the early 1970s, because TEL quickly damaged the new catalytic convertors by coating the platinum catalyst. Leaded gasoline similarly damages oxygen sensors on later model cars. Lead content began phasing down in 1975, with premium leaded gasoline disappearing from the market most places in 1981; leaded automotive street gasoline was eliminated from the U.S. market at the end of 1995. Today, lead is only found in avgas and racing gasoline for off-road use only.

In an effort to maintain octane levels near the acceptable 87 level, refiners began blending light end components, ethers, and alcohols into gasoline, which created new problems of hot drivability and hot restart, due to increased volatility. In 1989, the U.S. Environmental Protection Agency (EPA) implemented the first fuel volatility regulations; phase II of these regulations reduces volatility still further.

Today, neither TEL or benzene are considered environmentally acceptable as octane boosters. Instead, alcohols, ethers, and MMT are used in gasoline mixtures with higher basic octane ratings. The 1990 Clean Air Act required compositional changes in gasolines in areas of the United States with the worst air pollution problems. These "oxygenated" fuel mandates began with certain cities and towns during the winter in Colorado, Nevada, Arizona, New Mexico, and Texas, and expanded in 1992 to include 39 areas of the country. Oxygenated fuels contain oxygen-bearing ethers or alcohols, which not only improve octane ratings but chemically lean the air-fuel mixture of older cars under all conditions, and of newer, oxygen sensor-equipped cars when operating at full throttle. The 1995 regulations require fuel composition changes to address ozone pollution in the worst areas. U.S. regulations in the 1990s not only deal with the environmental impact of gasoline, but at the same time, American Society of Testing and Materials (ASTM) regulations provide specifications and guidelines to control performance characteristics of gasoline. The maximum legal quantity of oxygen in street gasoline is 2.7 percent, as occurs in a 15 percent blend of MTBE.

Gasoline and Air

Gasoline is a mixture of hydrocarbons with about 15 percent hydrogen and 85 percent carbon by weight.

Raw liquid gasoline will not burn. It must be vaporized and mixed with sufficient air to provide enough oxygen for combustion.

Air is a mixture of 21 percent oxygen, 78 percent nitrogen, and 1 percent other gases. Only the oxygen combines with the gasoline; the rest is along for the ride. For efficient combustion, gasoline must be atomized—broken up into tiny particles—and well mixed with the air.

The same air-gasoline mixture will not work well under all circumstances; the mixture must be adjusted constantly according to speed, load, and temperature, which affect flame speed, engine volumetric efficiency, and gasoline vaporization characteristics.

The chemically ideal static air-fuel mixture by weight, in which all air and gasoline are consumed, occurs with approximately 14.6 parts air and 1 part fuel, assuming there is time for complete combustion. The rich burn limit for an engine at normal operating temperature is about 6.0 to 1; the lean limit (for an EFI engine) is above 22 to 1. The following table, courtesy of Edelbrock, indicates characteristics of various mixtures. These figures do not indicate anything about the effect of various mixtures on exhaust emission.

Oxygenate gasoline additive limits.

	A (% vol)	B (% vol)
Methanol, suitable stabilizing agents must be added [a]	3%	3%
Ethanol, stabilizing agents may be necessary [a]	5%	5%
Isopropyl alcohol	5%	10%
TBA	7%	7%
Isobutyl alcohol	7%	10%
Ethers containing 5 or more carbon atoms per molecule [a]	10%	15%
Other organic oxygenates defined in Annex section I	7%	10%
Mixture of any organic oxygenates defined [b] Annex section I	2.5% oxygen weight, not exceeding the individual limits fixed above for each component	3.7% oxygen weight, not exceeding the individual limits fixed above for each component

Notes:
(a) In accordance with national specifications or, where these do not exist, industry specifications
(b) Acetone is authorized up to 0.8% by volume when it is present as a by-product of the manufacture of certain organic oxygenate compounds
(c) Not all countries permit levels exceeding those in column (A) even if the pump is labeled

AFR	Comment
6.0	Rich Burn Limit (fully warm engine)
9.0	Black smoke/low power
11.5	Approximate Rich Best Torque at wide open throttle
12.2	Safe Best Power at wide open throttle
13.3	Approximate Lean Best Torque
14.6	Stoichiometric AFR (chemically ideal)
15.5	Lean Cruise
16.5	Usual Best economy
18.0	Carburetted Lean Burn Limit
22+	EEC/EFI Lean Burn Limit

Gasoline Composition

Gasolines are blends of various light hydrocarbon ring or chain molecules. They fall into four families—aromatics, olefins, paraffins or saturates, and cycloparaffins or naphthenes. The density and energy content of any gasoline is therefore a function of the density and BTU content of the

Approximate Octane Blending Values		
	Blending RON	**Blending MON**
Methanol	127 - 136	99 - 104
Ethanol	120 - 135	100 - 106
Tert. butanol	104 - 110	90 - 98
Methanol/TBA 50/50	115 - 123	96 - 104
MTBE	115 - 123	98 - 105
TAME	111 - 116	98 - 103

particular ingredients.

Gasoline generally contains aromatics and paraffins, also called saturates, since they are saturated with the maximum amount of hydrogen possible. Olefins are common in street gasoline, and are high energy, but are not typically found in racing gas. Aromatics, based on the benzene ring molecule, tend to have a high energy content per gallon because they are densely packed and heavy per gallon, although this changes with temperature. They have excellent antiknock capabilities and are very high octane blending components. They tend to enhance the BTU content of a gasoline.

Olefins, with the general formula C_nH_{2n}, are unsaturated hydrocarbons, such as ethylene, and are characterized by relatively great chemical activity. They are the easiest hydrocarbons to break apart and combust, and are therefore susceptible to auto-ignition and detonation. They are found in low-octane street gasolines, but rarely in avgas or high-octane racing gasolines. They are not saturated with hydrogen and therefore have some double valence bonds. Olefinic hydrocarbons include ethylene (C_2H_4).

The paraffin series of hydrocarbons is a homologous group of saturated aliphatic hydrocarbons having the general formula $C_nH_{2(n+2)}$, the simplest and most abundant of which is methane. Other light paraffins include ethane (C_2H_6), butane (C_4H_{10}), iso-pentane (C_5H_{12}), and iso-octane (C_8H_{18}).

Aromatics are hydrocarbons that contain the six carbon ring characteristic of the benzene series and related organic groups. Examples of aromatics are toluene ($CH_3C_6H_5$), benzene (C_6H_6), and xylene $C_6H_4(CH_3)_2$. Aromatic hydrocarbons are much more densely packed than straight-line (chain) hydrocarbons, which means there are more BTUs per gallon.

Early gasoline was a light distillate made from crude oil, in which the aromatic level depended purely on what was present in the crude (thought to be an average of 9 percent by volume). Postwar aromatic content is estimated at an average of 12–14 percent by volume, with modern aromatic content of gasolines varying from 12–65 percent by volume. Aromatic content is lower in winter because of the dilution effect of volatile, nonaromatic hydrocarbons like butane. Aromatic content has decreased with the post-1986 introduction of oxygenates, which dilute aromatic content and provide octane enhancement once provided by aromatics.

Gasoline Additives

Gasoline additives are evaluated in terms of the physical and chemical properties of corrosion/water tolerance, materials compatibility, binding and handling, and effects on fuel quality. In terms of engine performance, additives are evaluated for carburetor deposit control and cleanup, fuel injector cleanliness, intake and exhaust manifold effects, and intake valve cleanliness. In terms of "no-harm" effects, additives are evaluated for octane requirement increase, intake valve sticking, ring sticking, bearing corrosion, oil thickening, and black sludge.

In the mid-1950s, refiners introduced the first detergents for effectively controlling deposit formation in the carburetor, consisting of amines, amino amides, and other low-weight nitrogen compounds. The low thermal stability of these compounds resulted in deposits at intake system hot spots. Refiners introduced second-generation higher molecular weight products, referred to as polymeric dispersants, in the 1970s. These additives are made up of polybutene amines, polybutene succinimides, poly-alkylene Mannichs, and polyether amines. The modern deposit additives are mixtures of dispersants and detergents, with an oleophilic tail and a polar head group.

Both are surface-active materials which are thought to function by the polar head's attraction to the metal surface, and the oleophilic tail projecting into the fluid, forming a protective film on the metal surface that prevents deposition of deposit precursors and particulates. Dispersants dissolve deposit precursors and keep them suspended.

Due to the fact that gasoline is a vapor in the hotter regions of the engine around intake valves, additives are often dissolved in a carrier oil which remains a liquid in temperatures reaching 400 degrees.

In general, gasoline additives include octane improvers, some of which, like TEL, are no longer street-legal in the U.S. Over-the-counter octane boosters, like VP's C5, will raise octane rating up to eight numbers, depending on what you start with and what other additives are already in the fuel, and are designed to provide increased performance—abnormal combustion, such as detonation, kills power. C5 will harm catalysts.

Detergents/deposit control additives eliminate or remove fuel system deposits. Fluidizer oils are used with deposit control additives to control intake valve deposits. Anti-icers prevent fuel-line freeze-up. Corrosion inhibitors minimize fuel system corrosion. Anti-oxidants minimize gum formation in stored gasoline. Metal deactivators minimize the effect of metal-based components that may exist in gasoline. And lead replacement additives minimize exhaust valve recession by replacing the protective lead coating, which leaded gasolines had previously provided on exhaust valves.

Many of these additives are

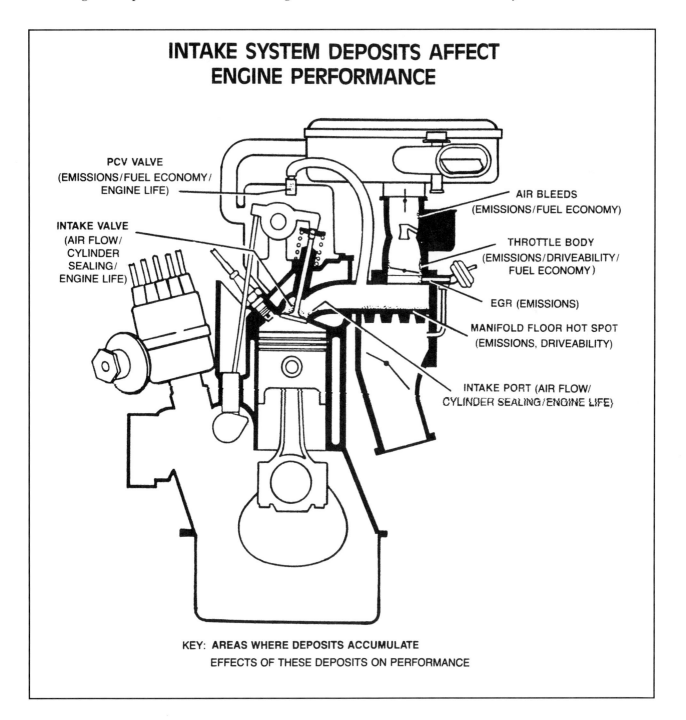

INTAKE SYSTEM DEPOSITS AFFECT ENGINE PERFORMANCE

KEY: AREAS WHERE DEPOSITS ACCUMULATE
EFFECTS OF THESE DEPOSITS ON PERFORMANCE

Unocal multifuel 327-cubic-inch Chevy test engine utilizes two carbs with entirely separate fuel supplies, each feeding four cylinders (1, 7, 4, 6) and (2, 8, 3, 5). Engine provides a dynamic comparision of different fuel formulations.

found in pump gasoline, and many are also available over-the-counter at auto parts stores in diluted form. These additives are blended into street gasoline in tiny amounts. For example, 100 pounds of deposit control additive treats 20,000 gallons of gasoline. Consumer benefits of additives include improved performance, increased engine life, lower deposits, drivability improvements, and better fuel economy. Additives are expensive, and refiners do not add them frivolously. Street gasolines are designed to comply with ASTM fuel standards guidelines, and additives are needed to comply. Most gasolines are also designed to meet BMW's test for the ability to keep valves clean.

Street Gasoline Quality Defined

In order to cold start engines easily, at least some components of gasoline must boil into a vapor at very low temperatures. As an engine warms up, gasoline should vaporize at an increasing rate for good drivability. In parts of the world with hot summers and cold winters, refiners change gasoline formulations with the seasons in order to provide very easy vaporization in cold weather and yet prevent vapor lock and pollution in hot weather.

For good drivability, street gasoline should have high anti-knock characteristics throughout the boiling range of its components. It should contain only minimal amounts of hydrocarbons with extremely high boiling points in order to produce excellent distribution under all conditions, and to prevent oil dilution and crankcase deposits. Gum content should be low in order to prevent valve sticking and deposits in the carburetor and intake tract. In order to prevent storage problems, street gasolines should have good stability against oxidation to prevent deterioration and gum formation in storage. Anti-oxidant chemicals are added to gasoline to protect against for-

Single cylinder variable compression ratio engine evaluates fuel octane by comparing trace knock conditions of a test fuel to that of reference fuels iso-octane and n-heptane, providing research or motor octane numbers. *Chevron*

mation of gum and peroxides. Today, phenylene diamine, aminophenols, and dibutyl-cresol have been replaced in this function by ortho-alkylated phenols.

Metal deactivators such as amine prevent trace amounts of copper picked up from piping or the vehicle's fuel system from acting as a catalyst in the formation of undesirable materials in gasoline. Alcohol de-icer additives prevent carburetor and fuel line freezing, and antirust additives prevent fuel system water from oxidizing metals in the fuel system. Detergent additives in gasoline keep carburetors and fuel injectors clean and fully functional. Phosphorous additives prevent spark plug and ignition fouling. Dye additives identify leaded gasoline.

Octane-Improving Additives

The difference between the two common octane rating methods (RON and MON) is called sensitivity. The octane rating displayed by law on all gas pumps is an average of the two methods ((R+M)/2), and this is probably the best overall antiknock performance indicator for street vehicles. Tetraethyl lead and lead scavenging additives have been added to street gasoline for many years, but in 1995 TEL was banned from street gasoline.

There is no advantage in using a higher octane fuel than needed to prevent detonation. High-octane fuel will not make more power or better gas mileage. As *Consumer Reports* once said, octane numbers are like shoe sizes. If you have size 8 feet, you won't run any faster in size 10 track shoes. Buying gasoline with a higher octane than your motor needs is a waste of money. Intermittent light knock is not usually significant in either causing engine damage or hurting power output. Heavy knock over longer periods can damage the engine, and severe knock can damage racing engines in seconds, as well as killing power.

Unleaded street gasoline is

1993 ZR-1 Chevrolet engine illustrates state-of-the-art in street gasoline engines. Engine produced 400 horsepower on 93 octane fuel using four cams and 32 valves.

usually available in the U.S. in 87, 89, and 92 octane blends. A few companies offer 93 octane super premium, and a few offer 91, 88, and 86 octanes, particularly in high altitude locations where motor octane requirements are lowered by decreased atmospheric pressure. A few gas stations around the country (in California, for example) sell 100 octane racing gasoline. In general, premium grade gasolines are more dense than regular octane fuel. Unleaded premium is substantially more dense than unleaded regular, and unleaded fuels are denser than leaded varieties because increased amounts of denser aromatics are needed to achieve equivalent unleaded octane ratings.

Volatility

Street gasoline volatility is defined by its distillation curve (the boiling range of its components) and by its vapor pressure, which is now prescribed by law. Temperatures required to boil 10, 50, and 90 percent of gasoline define its distillation curve. Street gasoline must boil easily in cold weather for starting, and is reformulated for warm weather in order to avoid vapor lock and emissions systems problems.

Reformulated and Oxygenated Gasolines

Oxygenates have been used as gasoline additives since the 1950s, first in the form of isopropyl (rubbing) alcohol for anti-

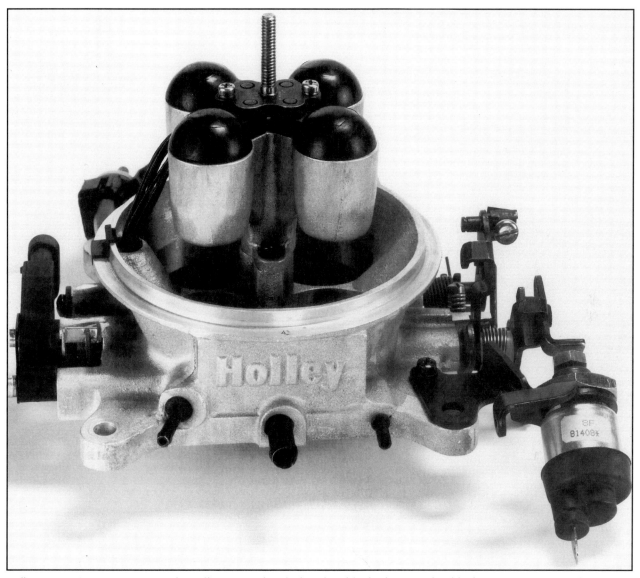

Holley Pro-Jection 4 system provides sufficient gasoline fueling for older high output big-block street-type engines being converted to fuel injection. Holley supplies optional closed-loop capability for lower speed mixture calibration with heated oxygen sensor. System lacks the inherent top-quality mixture distribution capabilities of port injection but is very cost effective.

icing. Arco "Clear" unleaded gasoline has contained gasoline-grade tertiary-butyl alcohol (GTBA) since 1969. In 1979, the Nebraska Gasohol committee began adding 10 percent ethanol by volume to gasoline and marketing the mixture as "gasohol" to stretch gasoline supplies during the oil embargo. In later years, in order to reduce air pollution, "reformulated" gasolines have been oxygenated by blending in ethanol, ethyl-tertiary-butyl-ether (ETBE), methyl-tertiary-butyl-ether (MTBE), tertiary anyl-methyl ether (TAME), and tertiary amyl-butyl ether (TABE). The bulk of oxygenates consists of MTBE and ethanol—MTBE is present in at least 25 percent of gasolines in the U.S.; ethanol in at least eight percent.

Again, these alcohols and ethers contain oxygen, perfectly mixed throughout the gasoline, which dissociates with the heat of combustion, providing additional oxygen to effectively lean mixtures in older cars and in newer cars operating at high throttle settings. Ethers and alcohols are blended with gasoline in a proportion to provide the legal limit of 2.7 percent oxygen, by weight. This corresponds to 10 percent ethanol, 15 percent MTBE, or 17.2 percent ETBE. MTBE and ETBE, made by reacting isobutylene with methanol or ethanol, eliminate certain problems with straight alcohols in fuel, such as methanol's corrosiveness to fuel supply systems and its hygroscopicity.

Benefits of oxygenates include reduced CO emissions, reduced ozone formation from other emissions, and reduced global warming compared to pure gasoline.

Other aspects of reformulated fuels are specified in the 1990 Clean Air Act. The Act provides for a maximum concentration of benzene of one percent by volume, a vapor pressure of 7.2 psi (or 8.1, depending on area), and no net increase in NO_x emissions.

From a performance point of view, oxygenates contain less energy than gasolines. A ten percent blend of ethanol and gasoline contains 3.4 percent less energy. Although there is less energy, the additional oxygen may result in more complete burning of the fuel. Older carbureted vehicles, which often ran excessively rich mixtures, will often show improved fuel economy from oxygenates. Late-model street vehicles with oxygen sensors will typically show a two percent loss of fuel economy, because the sensor will detect uncombusted oxygen in exhaust gases and richen the mixture to achieve stoichiometric ratios. Oxygenated fuels, with slightly greater specific energy, could theoretically allow richer mixtures than pure gasolines for more power. Adding oxygenates to gasoline tends to improve octane, although the base octane of gasoline stocks is adjusted by refiners in order to achieve target octane ratings for the complete fuel following oxygenation.

Detergents

Refiners have been adding anti-deposit detergents to gasoline since the 1950s to keep carbs clean. In the later 1970s and early 1980s, when increasing numbers of engines were available with port fuel injection, better detergents were required to prevent pintle injectors (with orifices as narrow as a human hair) from plugging and producing rough running engines. Port injection systems are also sensitive to intake valve deposits. The porous surfaces of coated intake valves absorbed enough fuel under some conditions to lean out the mixture, causing stumbling and stalling, especially during warm-up. Late model engines, idling at stoichiometric or leaner mixtures, or during cold running, are sensitive to anything that leans the mixture, even temporarily. It turned out the detergents to keep injectors clean could actually contribute to valve deposits.

Modern unleaded fuels containing more cracked components and oxygenated supplements contributed to the need for deposit control additives. Higher octane olefinic compounds, increasingly required to maintain octane as lead went away, and produced by cracking, are more prone to oxidation than previous gasoline components, and have been linked to deposits in critical fuel metering areas of carb jets and injectors and on intake valves. Heavy aromatic compounds, produced by severe refinery processing, contribute to deposits on combustion chambers and piston crowns, which increases octane requirement. MTBE has been shown to increase intake deposits in some engines. Older carburetted engines, with exhaust heated intake manifolds, are prone to intake manifold hot spot deposits. These cause drivability problems such as hesitation and stalling, decreased fuel economy, and increased exhaust emissions at idle and light cruise, and hurt wide-open throttle power by restricting airflow.

The BMW intake valve deposits test measures a gasoline's ability to prevent intake valve deposits. In the test, technicians disassemble a new or clean engine and weigh the intake valves. Following reassembly, the engine is run for 10,000 miles on a test track, then disassembled again and the valves weighed a second time. If deposits average less than 100 milligrams per valve, the gasoline tested passes with a rating of "Unlimited Mileage." If the deposits are more than 100 milligrams but less than 250, the gasoline passes with a "50,000 mile" rating. The BMW test is expensive. Following the 1990 Clean Air Act, all gasolines sold after January 1, 1995, must contain a deposit control additive.

6

Aviation Gasolines

Power-to-weight ratio has always been extremely important in aviation. In fact, the only engine parameter more important is reliability. The combination of these two requirements led early on in the history of aviation to high compression ratios and special high-octane aviation gasolines that were made to exacting standards for higher octane, lower Reid Vapor Pressure (RVP), and low moisture content. Later standards regulated levels of lead content, reaction with elastomers in the engine and fuel system, and other parameters.

Following WWI, GM patented a blend of 80:20 cyclohexane and benzol for aviation use. Phillips 66 (whose avgas products were at one time called "Phillips 77 Aviation") also began producing aviation petroleum products early on, and sponsored aircraft in competition such as the 1927 $25,000 Dole Pineapple Race from Oakland to Honolulu (Winning time for the 2,437-mile flight was 26 hours, 17 minutes, and 33 seconds). Phillips also sponsored Wiley Post in his 1934 flight nearly 55,000 feet into the stratosphere.

Research breakthroughs in the 1930s, led to development of avgas with an octane rating of 100 without tetraethyl lead. This provided excellent antiknock properties permitting high compression ratios and consequent higher horsepower, as well as providing superior energy content for quicker liftoff, increased maneuverability, and improved range for military aircraft in WWII. One hundred octane aviation gasoline powered Spitfires in the Battle of Britain. With the addition of lead, 100 octane fuel became 115 octane, which permitted even higher power levels.

During the war, Phillips and other oil companies subordinated all other interests to the war effort, and this period was a watershed for the development of extremely

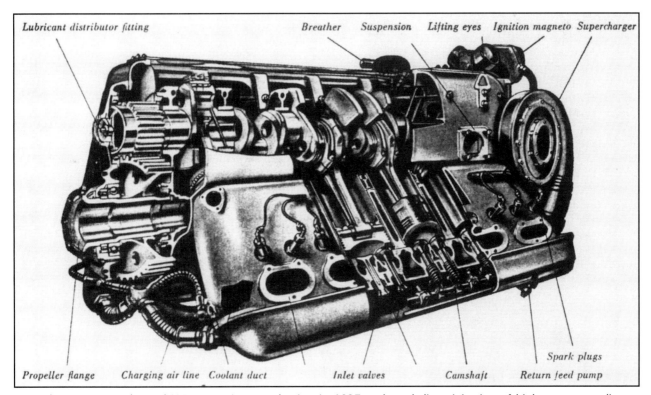

Mercedes-Benz supercharged V-12 went into production in 1937 and used direct-injection of high-octane gasoline to achieve 2 horsepower per kg in Luftwaffe fighter planes in WWII.

Lyle Sheldon's *Rare Bear* competes in unlimited class pylon air races at Reno and Phoenix achieving speeds comparable to commercial subsonic jet aircraft—at 500 feet above the desert, with 50-year-old radial piston engines! Planes like this use the highest possible octane fuel and are over-boosted to extreme power levels never seen in WWII combat.

P-51 Mustangs are the mainstay of modern unlimited class air racing. Rolls Royce V-12s are pumped to power levels as high as twice or more the stock power ratings.

Above and right
The *Pond Racer*, built by Burt Rutan's Scales Composites, was an attempt to build a modern aircraft competitive with the unlimited class warbirds, which would prevent the attrition of old planes and engines under the harsh conditions of air racing. The plane used twin Nissan 300 ZX-type automotive racing engines with under-driven propellers. Plane reportedly used oxygenated fuels, extreme levels of turbo boost, and engine layouts so constrained that technicians had to cool the engines following shutdown with leaf-blowers to protect the cowlings from heat damage. Plane suffered teething problems (such as the pictured aftermath of an in-flight fire) over a number of years as Rutan closed in on Gold-class speeds. Unfortunately, the plane was destroyed in a crash, and the effort abandoned after millions had been spent.

high-performance gasolines. Many blends of hydrocarbons, synthetics, and alcohols were tried as avgas during WWII when fuel was scare, performance everything, and cost only secondary. Highly leaded branched paraffins like triptane achieved very high octane ratings, which enabled tremendously high levels of supercharged manifold pressure. Unfortunately, fuels like triptane have such high production costs that expense is a problem for civilian use, even in racing. Research in polymerization and alkylation led to high octane products from these more intensive manufacturing methods. Using "flying laboratories," all sorts of synthetic and natural hydrocarbons were tested under every imaginable condition, from arctic to sun-blasted desert. Much of what is known today about high-performance racing and aviation gasolines is based on war-era research.

Following the war, there was a rapid shift of military aircraft to jet-engine technology (while armored vehicles shifted to diesels), and a loss of interest by the military in high-performance gasolines. Aviation gasolines were increasingly used for civilian general aviation in the United States.

As early as 1946, experts began suggesting the possibility of using auto gas in aircraft, but it was not until the 1970s that the Experimental Aircraft Association decided to test auto gas in planes for Supplemental Type Certificates, to legally use auto gas for specific aircraft. The test involves a flight to high altitude with preheated fuel in the tanks. (Airframe and engine manufacturers, perhaps constrained by the fear of lawsuits, have strictly forbidden the use of auto gas in their products, although by the early 1990s the EAA and Fixed Base Operator Charles Petersen had distributed auto gas STCs for 9,000 aircraft.)

Another P-51 at Reno races with cowling removed. This modified engine requires 120-160 octane fuel.

Circus Circus offshore racing boat uses twin automotive-type engines and surface drives to fly over the water. Marine conditions are similar to aviation in that fuel can essentially be formulated for constant-speed engines where transitional response is not nearly as important as in automotive racers. High-octane gasolines are used in turbo-supercharged engines that achieve power levels as high as 1,500 horsepower.

In the meantime, radial engines almost disappeared, and by the peak manufacturing years for general aviation aircraft in the late 1970s, engines were mostly four- and six-cylinder horizontally opposed air-cooled powerplants from Lycoming and Continental.

Although avgas was available through the 1940s and 1950s in 80 octane (dyed red), 91/96, 100LL (blue), 100/130 (green), and 115/145 (purple). The 115/145 and 91/96 disappeared by the late 1960s in the U.S. as the cost of distribution rose out of proportion to the size of the market, followed by 80 octane, which became nearly extinct in the U.S. due to relative lack of demand by about 1980. Eventually 100/130 also became virtually extinct, due to the Environmental Protection Agencies actions against lead in the atmosphere.

The rating of aviation fuels has continued to be based on the original fuels testing for high-performance WWII aircraft, rather than the automotive test procedures for octane. The tests, which yield two "performance" numbers (100/130, for example), were designed around the fact that all piston-engine aircraft have pilot-adjustable air-fuel mixture controls. The first number is a "lean mixture method" (ASTM D 2700), which tests the fuel's antiknock capacity at stoichiometric mixtures. The

second number is based on the "supercharge method" (ASTM D 909), in which antiknock capacity is based on the considerably rich mixtures used on supercharged or turbocharged heavy aircraft on climb-out under boost conditions. These aircraft sometimes used water and alcohol injection under these conditions to fight knock and allow higher levels of boost. Unocal fuels engineer Tim Wusz points out that avgas is typically blended from highly branched straight-cut alkylates, which have little or no sensitivity. Iso-octane is typical—its motor and research octane are the same. Some even have negative sensitivity. Comparing avgas performance numbers to automotive gas octane is not entirely straightforward. Unocal's 76 Racing Gasoline, for example, has a motor octane of 104, an (R+M)/2 octane of 108, and aviation performance numbers of 112/160. Wusz says a rule of thumb is to think of the first number of the avgas as the motor octane, which is clearly true in the case of the common modern American avgas 100LL.

General Aviation suffered a steep decline in the 1980s, and by 1995 Shell and certain other suppliers had abandoned the general aviation market, concentrating only on supplying jet fuel. This left the major players—Phillips 66, Chevron, Exxon, and Texaco—as avgas suppliers. Phillips, in particular, chose to expand its avgas business. By 1993 Phillips 66 aviation was selling 22,000 barrels of fuel per day, of which 11,000 barrels (both avgas and Jet-A) went to general aviation,

6,000 to airlines, and 3,000 to the military.

100LL is still widely available, but the trend, particularly in older, lower performance aircraft which once used 80/87 or 91/96 octane, is toward the use of premium auto gas, due to the nearly 100 percent price premium on avgas. At the same time, the federal Clean Air Act of 1990 has been forcing the replacement of many old tank farms with newer, more environmentally acceptable systems, making it less attractive for aviation FBOs and other suppliers to maintain tank systems.

Avgas and Auto Gas

In the 1990s avgas is often used as a base blending agent for racing gasolines, while auto gas is increasingly used to fuel aircraft, which are relatively insensitive to octane rating. There are some interesting implications. A Cessna Skylane, filled with 100/130, contains about a pound of metallic lead, plus a quantity of the ethylene bromide used to scavenge the lead. (115/145 used up to 1.28 grams of lead/L, while 100/130 used up to 0.85 grams. 100LL used up to 0.56 grams lead/L.) Particularly while taxiing, plugs frequently fouled with lead, which would often be masked by the twin-plug engine design, only to appear during run-up, when the engine was operated individually on the right and left magnetos. With the phase-out of 80/87 and the switch to 100LL, there was some expectation of problems with lead-fouling, but engine-maker Continental had not found this to be the case in engines originally designed for 80/87. However, excessive lead has been blamed for valve problems in low-compression engines where the valve materials and low compression permit chemicals in leaded

Left and above
The "lesser" air racing classes provide the opportunity for serious innovation in airframe and engine design. Designers frequently use high output automotive engines with sophisticated electronic engine management not seen in the unlimited-class warbirds.

Above and right
Supercharged lake boat uses mechanical fuel injection and very high-octane aviation-type gasoline fuel. Fuels like this are mostly leaded branched paraffins.

avgas to erode valve faces and stems. High-performance aviation engines with higher combustion temperatures, well-calibrated injection systems, and sodium-filled Inconel exhaust valves have been less susceptible to lead problems than lower performance carburetted aviation engines.

Today, aviation fuel is mainly alkylate—branched-chain hydrocarbons formed by combining isoparaffins and olefins in the presence of a strong acidic catalyst. "Aviation fuel," said Richard Collins in an article in *Flying* magazine on the use of auto gas in planes, "is strong on paraffins; car gas is strong on aromatics, which do not have the resistance to pre-ignition required of fuel used in aircraft engines." Collins pointed out that pre-ignition occurs when the air-fuel mixture lights off early from a hot spot in the combustion chamber, such as a sharp metal edge, glowing carbon deposit, or red-hot spark plug electrode, and cylinder pressures can increase from a normal peak of 900 psi to 2,000 psi, while head temperatures rise up to 20 degrees C per second. The result can be a holed piston or a cylinder coming off.

The distillation curve of avgas is typically different from auto gas, and the RVP is lower—7 psi vs auto gas' 9–15 psi, which varies seasonally. Since avgas therefore reacts differently than auto gas to temperature changes, mixture distribution to various engine cylinders might be affected by a change to auto gas and could potentially cause disastrously lean mixtures in some cylinders, which could facilitate pre-ignition and detonation. Aromatics in auto gas allow it to hold up to four times as much dissolved water as avgas. That water could release at high altitude, meaning that some high-flying aircraft would require anti-icing additives with high aromatic fuel. In Europe, where avgas contains slightly higher aromatic levels than in the U.S., the Cessna 421 requires such additives.

The relatively low octane of auto gas appears to be manageable in low- or medium-compression naturally aspirated aircraft. Both service experience and test-stand

data indicate all low-compression engines originally requiring 80/87 are fine with auto gas, as are engines like the Lycoming O-540 that originally required 91/96, but which automatically became 100/130 engines with its phase-out. Not only did these engines not have problems with detonation, but auto gas did not produce any unusual deposits, wear, or corrosion. The consensus of experts seems to be to draw the line at turbocharged aircraft engines, which require more precise management of mixtures, manifold pressures, EGT, and engine cooling. In fact, racing aircraft with high manifold pressures are often running fuels higher in octane than 100LL, for power and longevity.

Other concerns with auto gas in planes include the possibility of chlorinated auto gas producing corrosion and the possibility of damage to noncompatible elastomers, such as carburetor floats sinking. Many of these floats have already been replaced with brass units. The EAA warns about the use of auto gases high in toluene attacking epoxy fuel tanks, but many composite experimental aircraft have always used auto gas, with no problems.

The most important incompatibility between auto gas and avgas is vapor pressure, or the susceptibility to boiling. With its higher RVP, auto gas has more tendency to vapor lock, since it is not designed for high and hot conditions as encountered inside the cowling of an air-cooled aircraft during cruise climb-out. Many states have not had gasoline quality control standards that would ensure known-quantity fuel desirable for airplane use. Unlike avgas, which has an RVP below 7, auto gas is formulated according to the climate, with an RVP as high as 15 in cold weather, to provide easy starting, but as low as 9 in hot weather, to avoid vapor lock. The RVP of a given batch of gas is not particularly predictable. Nor is the requirement of a particular aircraft on a particular day. Unfortunately, aircraft can easily make it from snow country to desert climates on the same gasoline. The routing of a fuel system, its bends and turns and sharp edges, restrictive filters, engine-driven fuel pumps, the ventilation capability of a particular aircraft, and the power level and angle of attack of a particular cruise setting or climb-out all affect the possibility of vapor lock in an aircraft. By the time fuel reaches the injectors or carb of an aircraft engine, pressure is down and the fuel might be quite warm—as hot as 200 degrees in the float chamber of a recently parked aircraft, according to tests. At some point, heated fuel will begin to form vapor bubbles, which can eventually restrict the amount of fuel arriving at the various cylinders. The effects can range from minor fluctuations in fuel pressure to a steady loss of pressure and rising exhaust gas and cylinder head temperatures. In extreme vapor lock, the engine could surge dramatically or even quit entirely. The worst case is likely to be a takeoff in a heat-soaked aircraft, which is a significantly different problem than problems at altitude.

Most people reading this book are likely to be more interested in the possibility of using high-octane avgas in high-performance automobiles rather than the reverse, as a cheaper source of high octane fuel. (100LL hovers near $2.00 a gallon, whereas racing fuels are likely to be twice that cost.) Aviation gasolines have often been blended with other compounds like toluene or xylene to produce high-octane racing fuels for motor cars. Racers considering using avgas should know that the taxing structures may make it illegal to just use avgas in street automobiles. In addition, as of 1995, the use of any leaded gasoline for highway use is a federal crime.

But there are other potential problems with avgas in high output cars. The distillation curve for aircraft is essentially designed for constant-speed engines, in which acceleration is not an issue, but avoiding vapor lock is all important. Good racing fuels have plenty of light end components to provide crisp light-off of combustion and prevent the mushy-feeling acceleration that can be a consequence of fuel with the wrong distillation curve. Fast, crisp engine response is essential for auto racers coming out of curves, and particularly for drag cars. The other problem with avgas is that its low RVP can and does make for harder starting. Pilots approach piston-engine aircraft in cold weather with trepidation. Aircraft engine block heaters and such are common for colder climates, and lots of cranking is typical on cold days, even in relatively cold climates.

There is still a demand for ultra high-performance aviation gasolines for racing aircraft. "Air Race Fuel," from VP hydrocarbons, with performance numbers of roughly 120/160, is the highest octane gasoline of any kind sold by VP, and has been used for some extremely high output forced-induction automotive engine applications.

7

Gasoline as a Racing Fuel

From the dawn of automotive racing, competitors tried home brew blends of various hydrocarbons, and oxygenated and nitrated hydrocarbon fuels in a quest for more power, sometimes destroying their engines along the way. WWII spurred massive research by the commercial oil companies into high-energy and high-octane aviation fuels, but the first commercially available racing gasoline for NASCAR stock car racing may have been produced by the Pure Oil Company in the mid-1950s. Pure (which later merged with Unocal) had begun to sponsor racers after executives read in the Chicago paper about a Pure Oil dealer who had won the 160-lap Daytona race in 1951. Pure's race fuel replaced earlier stock car racer gasolines, which were nothing more than premium octane pump gas.

The earliest stock car racers made 100 horsepower from 7:1 compression ratios and did fine on pump gas. The only changes required in constituting gasoline for the earliest racers was to lower RVP in order to prevent vapor lock. To eliminate the need for carb adjustment when racing, early racing gasoline was similar in specific gravity to street gasoline, but by the late 1950s, octane had increased significantly.

Unocal racing fuels engineers were routinely testing racing gasoline octane using not Motor or Research but Road Octane Number (RdON), which is the antiknock performance of the fuel when subjected to real conditions found in multicylinder engines in automobiles. Pure Oil RdON testing led directly to a new 96.6 octane racing gasoline composed of 76 percent Fluid Cat Cracked stock and 24 percent alkylate, and containing two grams of lead per gallon. In its final version, Pure's race gas consisted of 3 percent isobutane, 5 percent light straight run, 41 percent alkylate, and 51 percent reformate, plus tetraethyl lead and an additional manganese octane improver additive. Pure Oil's "Tri-tane" additive package was tested in 1961, and found to have no problems resulting from the additive package of carburetor detergent, anti-icing additive, and phosphorus (tri-cresyl phosphate). Phosphorus, originally added to control surface ignition and rumble, was later discovered to have dubious benefits, but has been retained in Unocal race gas into the 1990s.

By the 1960s engine displacements had grown to over 400 cubic inches, and compression ratios in showroom stock cars were as high as 11:1. "Street" vehicles like the 435 horsepower L-88 Corvette carried a dashboard placard requiring 103 octane (RM) fuel, which could be obtained at service stations selling Sunoco 260 gasoline in the east, or, in the west, Chevron "White Pump" gas. Both these super premium fuels had motor octane of 94–95, and an (R+M)/2 of 98–99, according to Tom Wusz, and 3.5 grams of lead per gallon. "Blending fuels was a piece of cake back then," says Wusz. "All you had to do was adjust the amount of lead according to the quality of the fuel stock to reach a target octane."

Raced on tracks that permitted speeds over 200 miles per hour, these big high output engines needed plenty of octane to satisfy antiknock requirements in large-bore, hemispherical head, high-compression engines. Pure/Unocal racing gasoline, which increased in octane in the 1950s from 87 to 98 ((R+M)/2), increased to 100 (RON of 107, MON of 95 or so) in the early 1960s, where it stayed until 1973. In the 1970s Unocal NASCAR race gas increased to 104, and in the late 1980s to 108.

In the late 1980s and 1990s, the extremely sensitive interplay between engine developments, race track conditions and sanctioning body regulations regarding legal fuel and engine specifications resulted in frequent changes in fuels formulation for sophisticated racing like Formula One. The need to quickly respond with new formulations has caused fuels engineers to bring to bear the most modern computer simulation and fuel testing facilities. Computers select potential fuel blends, evaluating the legality of the components and their cost and benefits as associated with changes in engine tune. The most promising are selected for lab testing, and the most promising of these are tested in race cars. A really perfect fuel blend can result in perhaps 50 additional horsepower in a Formula One car.

Racing Gasoline

Gasoline has always been the most widely used piston-engine racing fuel because it's easily available and has vaporization properties that make it easily metered and uniformly distributed to the cylinders of racing engines of various configurations by inexpensive carburetors. In addition to these useful characteristics, gasoline has ignition delay characteristics that make it useful as a racing fuel. Dynamometer tests have been

55

run extensively using racing gasolines and are increasingly published and available to the public. Racers are familiar with the fuel and know what to expect.

Gasoline's biggest problems as a high-performance fuel are its relatively low specific energy and its tendency to knock at high compression ratios. Gasoline's heating value is high—higher than nitromethane, alcohol, and the other high-performance fuels—but it needs a lot of air to burn, which makes engine volumetric efficiency a limiting factor when making power with gasolines. Gasoline's unfortunate tendency to spontaneously pre-ignite in advance of the expanding flame front under high output conditions produces dangerously rapid rates of combustion chamber pressure increase, exactly as if the piston crown is being slammed with a sledgehammer. This can crack pistons, break rings, and overload bearing surfaces. In addition to possible engine damage, power drops off under knocking conditions, in part due to increased heat losses into the combustion chamber surfaces due to violent turbulence.

A related measure of combustion is the ignition temperature—the temperature at which the fuel components dissociate and react with the oxygen in air. Generally, the lower the ignition temperature, the lower the octane, although this is not true with aromatics.

Octane

Octane rating is usually considered the chief factor in evaluating gasolines for racing use. Clearly, motor octane number, with its varying and higher speed test parameters, is more relevant than RON in gauging a fuel's racing antiknock capabilities. According to Tim Wusz, racers should never care about RON, only MON, which is a much better antiknock indicator for the harsh conditions of racing. Some racing fuel suppliers use research octane number (RON) because it is often higher and easier to come by. According to ASTM method number D2699-81, the highest octane rating possible for the research and motor octane engine is 120.3, although pure toluene with large amounts of tetraethyl lead has been described as having an octane rating of 124. According to tests run for *Circle Track* magazine, for most brands of racing gasoline the actual octane in most cases exceeded the advertised octane ratings.

There are trade-offs in achieving the highest octane ratings. The highest octane fuels may have slightly less energy content per gallon, due to the use of aromatic components that are one to three percent less dense. Assuming an equal

MINIMUM OCTANE RATING
(R+M)/2 METHOD
116

This is about as good as it gets. Highly leaded blends of aromatics and branched paraffins achieve octane ratings in the 116 range, with RON levels in the 120 range and MON above 110 on up to 5 grams of tetra ethyl or tetra methyl lead.

state of tune, the lower octane fuel could make a few percent more torque and could result in a slightly richer air-fuel mixture (which is, after all, a matter of the weight of fuel and air), although in practice this difference does not appear to be measurable.

Flame Speed

Another important consideration for racing gasoline is burning speed, which is the speed at which a fuel releases its energy, measured not in crank rotation, but in milliseconds. In order to contribute to peak cylinder pressure, fuel must release all its energy by 20 degrees after top dead center.

A further consideration in racing fuels is heating or energy value, often specified in British Thermal Units (BTU) per pound of fuel (not per gallon, which is important, since air-fuel ratios are according to weight, not volume). Air and fuel density vary according to weather conditions, which can be a problem since carbs and injection meter fuel according to volume, and which is one reason why racers are interested in fuel density. One BTU is the amount of heat it takes to raise the temperature of one pound of water by one degree Fahrenheit.

Tests indicate racing gasolines vary in energy content by about 2 percent among the various samples tested. The ability of an engine to release all the fuel energy in the available time has a much bigger effect on power than actual BTU per pound. Both the aromatic hydrocarbon toluene, a frequent component of high-octane racing gas, and diesel fuel have very high energy content, but it would be impossible to release all the energy of, say, diesel fuel in a racing spark-ignition engine, since the ability to release this energy is inversely proportional to the vaporization temperature of the fuel components. Diesel boils at over 400 degrees, while toluene boils at 231 degrees.

Fuel heating value is sometimes confused with exhaust gas temperature. Experts say it is very doubtful that any of the common high-octane racing fuels would have a significant effect on EGT, but in any case, it is not possible to state in advance what the ideal EGT is for a particular engine without extensive testing on a dyno. Ideal peak EGT is affected greatly by the engine design and the fuel system and the air-fuel ratios it produces. Piston aircraft engines, with manually adjustable air-fuel mixture, are typically adjusted according to peak EGT, but best power or best economy are not the same as best safe power or economy, and aircraft engine manufacturers know this. On some engines, it may be dangerous to operate at peak EGT for best power. And the gasoline brand has essentially nothing to do with any of this.

Evaporative Cooling Effect

Another consideration for racing fuels is its cooling effect, which is a function of the fuel's heat of vaporization. A fuel's cooling effect is twofold: It can result in denser mixtures entering the cylinders, making more power, and it can lower peak combustion temperatures, fighting knock and potentially allowing a higher state of tune. The second, "supercharge method," number in an avgas performance number, for example, reflects the combustion cooling effect of extremely rich mixtures.

Specific Gravity

Specific gravity is the ratio of the mass of fuel (or any other solid or liquid) to the mass of an equal volume of distilled water at 4 degrees C (39.4 degrees F); gasolines with higher specific gravity are more dense. Greater density may mean more energy, but the reverse is sometimes true, which is why racers can't jet carbs for the same energy content by jetting for equal density, and why BTU content per gallon specs are important. If you are actually trying to jet for an equal amount of fuel energy entering the engine under different ambient conditions, it is necessary to multiply BTUs per pound by fuel density, in pounds per gallon.

Distillation Curve

A good distillation curve is vital for good transitional throttle response. Refiners determine a fuel distillation curve by heating the fuel and measuring the temperatures at which successive 10-percent segments of the fuel vaporize. The lighter the molecular weight, the lower the temperature at which a fuel will vaporize or boil. A good fuel vaporizes smoothly and

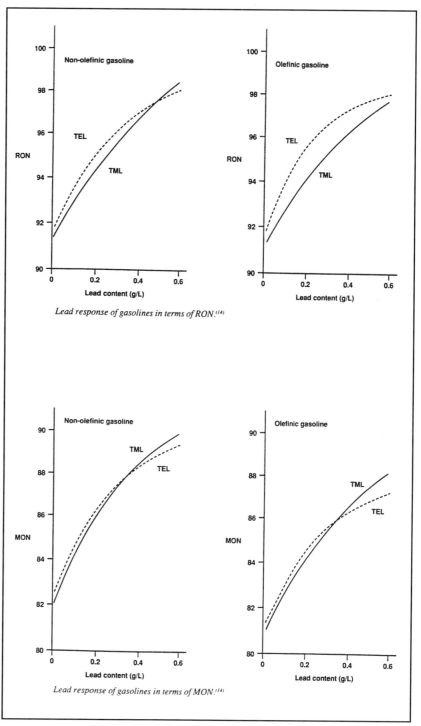

Lead response of gasolines in terms of RON.[14]

Lead response of gasolines in terms of MON.[14]

Olefinic and saturate gasoline components respond slightly differently to TEL and TML leading. Olefins provide high energy but lower octane quality, since the double bond is less stable and breaks down more easily. *SAE*

steadily as temperatures increase steadily up to 300 or so, at which time all the fuel should have vaporized. (Hydrocarbon fuels with components vaporizing above 400 degrees are not gasolines, but are considered diesel fuels.) A poorly constituted fuel might vaporize suddenly at particular temperatures. This is undesirable, since it might very well occur in the combustion chamber while some of the air-fuel mixture is already burning, affecting combustion and therefore power output. Since the boiling process takes a finite amount of time, fuels with broader distillation curves seem to have crisper performance. A fuel distillation curve which starts at low temperatures might vapor lock on hot days while waiting for a race to start, while a distillation curve which starts too high might be very difficult to start on cold days. The upper end of the distillation curve should be relatively low, so that all the fuel vaporizes and can be used in combustion. Fuels that vaporize well translate directly to more complete combustion and available power. By mixing fuel components with different boiling points, such as xylene (270 degrees), toluene (231 degrees), isopentane (82 degrees), and butane (31 degrees), gasoline is customized for a particular use. Various brands of fuel do have radically different distillation curves, which definitely affects the way an engine drives and makes power under various conditions. Racers having drivability problems with a particular brand of gas in a certain vehicle should try other brands with different distillation curves. According to Richard Riley of Phillips, "A flat spot in the distillation curve might give you a flat spot in your drivability curve." A different brand than the one you've been using might cure problems.

Racing Gasoline Changes, 1962-1977				
Component, %	1962	1970	1972	1977
Alkylate	41	51	61	55
Light St. Run	5	13	10	8
Isobutane	3	3		
Reformate	51	18	3	3
Xylene		15	26	30
n-Butane				4
RON minimum	103	103	103.5	107
MON minimum	97	97	98	100

Racing gasoline changes from 1962 to 1977. Composition changed and octane rating increased considerably, even as street gasoline octane peaked and began a downward trend around 1971. *Unocal*

Unleaded Racing Gasoline

Virtually all racing gas producers now offer unleaded racing gasolines, and these have conclusively been proven to resist knock as well as highly leaded racing gasolines, and to make as much or more power. The main advantage of lead has been that a certain octane-rated gasoline could be manufactured more cheaply by artificially improving the octane rating of lower octane fuel stocks with lead additives than by formulating a high-octane fuel with high-octane hydrocarbons. Not only that, when lead was available, gasoline blends of varying octane could be put within specs by simply varying the amount of lead to achieve the target octane at the end of the production process.

Power

Do all gasolines make equal power, or are some better than others? Dr. Bill Mallett, chemist and staff consultant at Unocal Fuels Technology Group, says, "Heavier gasolines have more power potential per gallon as measured in BTUs. However, to see a horsepower improvement in practice, well, it is really tough to show better acceleration or power." Exxon Research

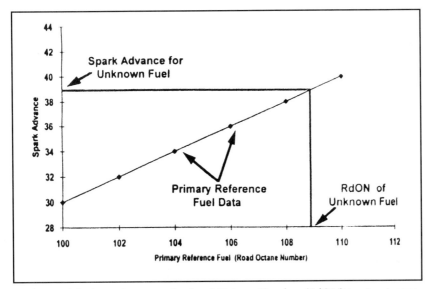

Engine and fuels engineers prefer Road Octane Number (RdON) to street gasoline ratings, because methods like the Modified Uniontown Method rate fuels in multicylinder engines installed in actual vehicles at realistic speed as opposed to the slow single cylinder test engines used for research and motor octane. *Unocal*

scientist Hugh Shannon points out that fuels can be blended with heavier molecules that get more power-producing compounds into an engine per volume of gasoline, which can be useful if fuel volume is limited. But remember, gasoline engines are air limited. And eliminating "light ends" from racing gas makes for hard starting and, some feel, poor transitional throttle performance (off and then back on the throttle, as in a turn) with even small changes in RVP. It seems clear that other factors tend to overwhelm the slightly varying heat content of various gasoline blends.

There are ways to blend gasolines to make more power. Smokey Yunick is quoted in *Circle Track* as saying, "I can take Unocal race fuel, for example, and add some things to it, make certain carb and ignition changes, and pick up 5–7 percent power. (Don't write or call about this; I won't tell you.)" The VP CU110 high-octane fuel tested by *Circle Track* in a dyno shootout of unleaded fuels outperformed leaded Turbo Blue. However, the VP CU110 consisted of almost 23 percent MTBE and almost 10 percent methanol, both of which are oxygen-bearing fuels with higher specific energy than gasoline, and would exceed the maximum of 2.7 percent oxygen legal for street gasoline. If a small amount of water became mixed with such a fuel, it would separate into two separate phases: Gasoline without methanol, and methanol with water. Mixing an exact known quantity of dyed water with fuel in a graduated cylinder and checking to see if the dyed mixture increases in volume after reaching equilibrium (as it would if methanol were present) is one method that sanctioning bodies have of testing for illegal "oxygen-bearing" components. Unocal and other oil companies have developed devices that measure the dielectric constant of a fuel and can detect as little as one percent nitro or alcohol in gasoline, even though the devises cannot tell what it is. Some

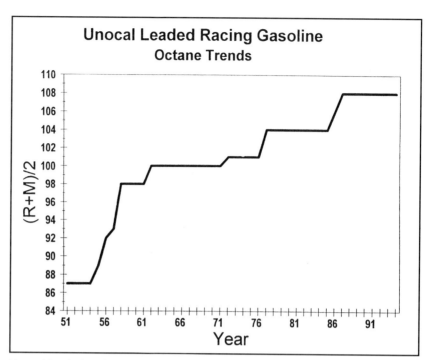

Unocal leaded racing gasoline octane trends. Pure Oil (later bought out by Unocal) introduced some of the first gasolines specifically designed for racing in the early 1950s. *Unocal*

racers had tried using masking agents to fool these detectors in order to fortify gasoline with oxygenated or nitrogenated fuels, but college professor Jeff Germain has developed a device that overcomes masking agents and can test for propylene oxide power boosters in gasoline. A gas chromatograph or capillary test can measure exactly what is in a sample of fuel, but these tests take several days and are not suitable for on-track testing.

Alcohol-gasoline mixtures will have higher specific energy than gasoline alone, but will only make more power if the mixture is richened, since the stoichiometric air-fuel mixture for alcohols is lower than gasoline. Without retuning, gasoline-alcohol mixtures will run lean and possibly make less power than straight gasoline. Gasoline can also be mixed with nitro propane, but again the mixture would need to be even more rich than alcohol-gasoline mixtures. "Nitro also drags down the octane of gasoline fast," says Tim Wusz.

Degradation

Racers working to produce maximum power in highly tuned engines need not only compensate for fuel and air density changes at race time but should be aware of the effect of sunlight, heat, and shelf life on stored gasoline. Some gasolines, such as Sunoco GT Plus unleaded, are photosensitive and should be stored in opaque containers or out of direct sunlight. While testing various unleaded fuels under dyno conditions, *Circle Track* magazine technicians saw a reddish sludge precipitate out of the fuel and settle on the bottom of a glass beaker. They were unable to get the reddish precipitate to remix back into the fuel. The TEL in leaded gasolines is also photosensitive.

Many fuels have additives that help prevent gum formation in stale gasoline, but racers storing fuel should know that the "light end" components of fuel—those that vaporize at the lowest temperatures and so provide easy starting and transitional throttle response—will vent from containers every time they are opened. It doesn't take very long for the escape of light end components to effect the distillation curve of gasolines. Even a week might be enough to make a detectable difference in the gasoline, according to Exxon's Shannon. The higher the temperature, the greater the tendency for gas to lose its light ends.

Because of the above, retailers selling racing gasoline out of 55-gallon drums could easily be a preferable source to one with larger tanks and relatively low volume, or distributors who are not preventing excessive venting of light end components. Some people have been known to store fuel drums upside down so that gasoline leaking out will be visible as a liquid rather than a colorless gas.

Real World Racing Gasoline Blends

Like aviation fuel, racing fuels currently consist of alkylates (branched, saturated isomers of paraffins), blended with smaller amounts of aromatic components. Tests by Unocal labs revealed that Sunoco GT Plus unleaded consists of 22 percent MTBE, 2 percent n-butane, 8.5 percent isopentane, 12.5 percent 2, 2, 4-trimethylpentane, and 42 percent toluene, plus over a hundred other hydrocarbons in tiny amounts totaling 13 percent. This amounts to 18 percent standard gasoline with additives. BTUs per gallon totaled roughly 125,000, and the octane rating ((R+M)/2) was 97.6.

Also according to these tests, VP CU110 consists of 22 percent MTBE, 9 percent methyl alcohol, 5 percent 2-methylhexane, 3.1 percent isopentane, 10.6 percent 2, 2, 4-trimethylpentane, 25 percent toluene, 16.2 percent m- and p-xylene, and roughly 9 percent of some 60 other hydrocarbons in tiny amounts. Essentially, this is 15 percent standard gasoline with high-octane additives.

MTBE and methanol are added to racing gasoline for their high specific energy and oxygen content and because they have very high octane ratings, although any "gasoline" with over 2.7 percent oxygen by weight has too much oxygen to be legal gasoline. Oxygenated components definitely affect target air-fuel ratios, and definitely require jetting or injection changes to make best power, and may also require retarded ignition timing. Isopentane is used for its high octane and because of its high vapor pressure. Butanes, pentanes, and hexanes are normal gasoline products of the catalytic cracking process. The aromatics have both a high heating value and a high octane, are somewhat dangerous, and can damage rubber and other fuel-system components.

The testing showed that the best unleaded gasolines were equal or better than leaded racing fuel, although certainly exotic mixtures like pure toluene and 5.0 grams of lead would produce stratospheric octanes, which would exceed the antiknock capability of any commercial racing gasoline. Some venders claim to have unleaded racing gasolines with antiknock capabilities good to nearly 17:1 compression. There are some exotic unleaded racing fuels, such as one from the French oil company Elf

Component	1951	1955	1956	1957	1958	1962	1969	1970	1972	1977	1985	1986	1987	1992
F.C.C. Stock				76							N/A	N/A	N/A	N/A
Alkylate	SERVICE STATION PREMIUM	7.0 RVP - S. S. PREMIUM	7.0 RVP - S. S. PREMIUM	24	41	41	PHOSPHORUS STUDY	51	61	55				
Light St. Run					5	5		13	10	8				
Isobutane					3	3		3						
Reformate					51	51		18	3	3				
Xylene								15	26	30				
n-Butane										4				
Isopentane														
Toluene														
Lead (GM/Gal)	~2 TEL	~2 TEL	~2 TEL	2.0 TEL	2.0 TEL	4.0 TML	4.0 TML	4.0 TML	4.0 TML	4.0 TML	4.1 TML	4.1 TEL	4.1 TEL	4.1 TEL
RON	91	94	97	98	100	103	103	103	103.5	107	107	110	112	112
MON	83	84	87	88	95	97	97	97	98	100	100	102	104	104
(R+M)/2	87	89	92	93	98	100	100	100	101	104	104	106	108	108

History of octane and components in Unocal racing gasolines. *Unocal*

which is a synthetic gasoline made from fermented beets.

Plug-reading during the dyno testing showed that spark plug fuel ring, center wire, side wire, and the shell all reacted the same as with leaded fuel. Detonation and oil on the ceramic looked the same as with leaded fuel.

Racing fuels are generally made in much smaller batches than street gasolines, and the resulting blends can vary by batch anywhere from a few tenths of one octane number to as much as five points. Stale fuel, particularly over 90 days old, can deteriorate enough that it's fair to say that racing gasoline has a shelf life. The aromatic capillary analysis process itself that reveals the ingredients has some margin of error. Unocal unleaded gasoline tested at RON of 106.7 in one anonymous lab test, 105.6 in another. The MON was 94.0 in one test, 92.9 in the other. The (R+M)/2 was therefore 100.4 in the first test, 99.3 in the second. Testing showed that some so-called unleaded racing fuels actually had illegally high levels of tetraethyl lead.

Formula One Race Fuel Blending

Formula One racing involves very large amounts of money being spent on extremely high output race cars. Given the potential for favorable publicity, oil companies such as Shell have worked closely with Formula One teams to optimize fuels for particular engines in a particular year. The experience of the McLaren team from 1988 to 1992 using Honda engines in Formula One racers is instructive.

The criteria used in fuels development included: 1) health and safety, 2) regulations, 3) reliability, and 4) performance. The performance criteria are complex. Does performance mean maximum horsepower? Maximum power to weight? Maximum power within a certain required fuel economy? How much reliability do you sacrifice for more power? How close to the edge of detonation will you run? How do you handle trade-offs between fuel octane, minimum spark advance for best torque, weight of fuel, density of fuel, compression ratio, turbo boost, etc?

1988

Turbochargers were still permitted in F1 racing in 1988, and the McLaren team used a V-6 Honda running 2.5 BAR boost. Given a 150-liter fuel tank, good volumetric fuel consumption was important. Past experience with the TAG-Porsche engine indicated that high-density aromatic fuel components would provide good fuel consumption. High flame speed was important given the 13.5K redline of the V-6 engine, which only allows 0.75 milliseconds for combustion. Sufficient volatility to produce crisp throttle response was essential, but a margin was required to prevent vapor lock. The TAG-Porsche engine had burned a blend of an aromatic component with premium gasoline added, in which the aromatic provided high research octane and good density (energy per volume),

Characteristics of MTBE and TAME as Racing Fuel Blend Components

	MTBE	TAME
Oxygen content, wt%	18.2	15.7
Boiling point, °C	55	86
Lower heat of combustion, MJ/kg	35.1	37.7
Heat of vaporization, MJ/kg	0.32	0.32
Stoichiometric air-fuel ratio	11.7:1	11.9:1
Blending RON (approx.)	115	111
Blending MON (approx.)	104	100

Octane Qualities of Some Pure Hydrocarbons

Compound	Formula	RON	MON
Isooctane	C_8H_{18}	100	100
Triptane	C_7H_{16}	112	101
Isodecane	$C_{10}H_{22}$	113	92
Cyclopentane	C_5H_{10}	101	85
Cyclohexane	C_6H_{12}	83	77
Toluene	C_7H_8	120	109
Xylene	C_8H_{10}	118	115

while the premium gasoline maintained good volatility and kept research octane within sanctioning body limits.

It turned out that the Honda engine ran better on an unleaded blend of an aromatic and just enough of a paraffin to bring the octane down to the legal limit. The heat, pressure, and turbulence of the turbocharger inlet system avoided potential volatility problems of the fuel blend, and tested power was 10 percent above that obtained with a commercially available premium gasoline. The aromatic yielded higher flame speed than isoparaffins, and provided enhanced performance despite its lower motor octane number.

1989

The 1989 F1 rules removed the turbo motor option, allowing only a 3.5-liter capacity normally aspirated motor, which would clearly require a different fuel composition for optimized operation. Honda responded with a V-10 engine. Although there was no fuel tank size requirement, good fuel economy was required to minimize weight. To run the highest possible compression ratio without detonation, maximum-allowed RON fuel octane (102) was essential, while maintaining knock margin. To insure legality throughout the range of fuel usage, candidate fuels were blended to maximum RON octane of 101.5. High flame speed was even more important with the loss of the turbo, a switch to 10 cylinders from six, and a slight increase in engine speed.

Given the entirely new engine configuration, engineers initially decided to abandon the unleaded aromatic-paraffin fuel blend for a more conventional gasoline, starting with a leaded avgas-type blend in which lead content allowed optimal spark advance. In the meantime, for street gasoline image reasons, Shell decided to develop an optimized unleaded fuel for the new engine, based on enhanced olefinic blending components. Olefins, which are not saturated with hydrogen, contain at least one high energy double bond between carbon atoms, and the chosen olefin was also highly dense. Shell selected a branched olefin component, blending it with an alkane (paraffin) to provide legal RON and improved volatility. The low motor octane of the fuel required spark advance to be retarded below MBT to avoid detonation, but the high energy of the fuel more than compensated for this power loss. The fuel was very dense and was ideal for races where fuel economy could be a problem.

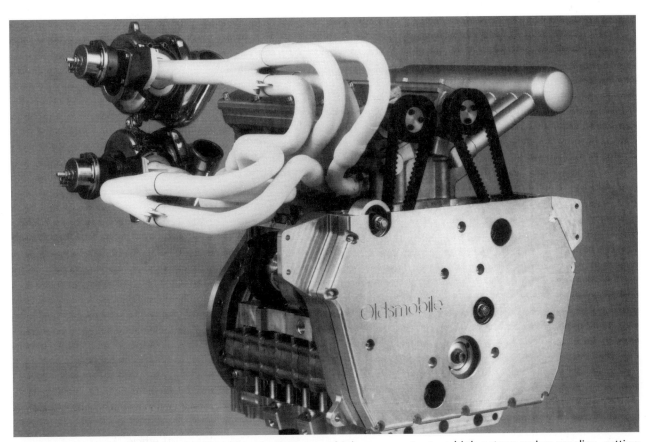

Oldsmobile's Quad 4 engine used twin turbos to achieve very high power output on high-octane racing gasoline, setting speed records at Fort Stockton well over 200 mph in the Aerotech car.

NHRA drag racing includes classes designed for high-octane racing gasoline engines, but blown and naturally aspirated.

1990

The same two fuels were used in 1990, with blend characteristics optimized such that one fuel maximized power, the other maximized fuel economy. An aromatic component was added to the "economy" fuel to further boost the fuel economy. The branched olefin fuel was used in all but two races in 1990.

Meanwhile, FIA had announced plans to eliminate leaded fuels for 1991, while Honda was developing a V-12 engine for the 1991 racing season, and Shell continued the search for high-power fuel components with high-octane and fast flame speeds, investigating certain high-power components developed in the 1960s such as "Shell dyne." These high-power components had high energy double bond or ring strained structures which were less stable and released more energy during combustion. Shell developed computer modeling tools to predict fuel characteristics, which showed that certain components had the potential to provide up to 6 percent more power than the branched olefin fuel, but unfortunately, they required enhancement with a high-octane, lower power component to reach the target of 101.5. Modeling initially assumed the volumetric efficiency of combustion is unaffected by fuel properties and that power potential was purely a function of fuel combustion with a unit mass of air, in equations that separately predict heat and moles of products versus reactants. Later models were more sophisticated. In spite of some doubts, Shell supplied the new fuels to Honda for testing.

1991

FIA reversed itself and allowed leaded fuels in 1991, but investment in unleaded fuels was not wasted: Unleaded fuels were competitive, and in any case, implementation of the lead ban was simply pushed back a year.

It turned out that Shell's high energy fuel components required complementary engine configurations due to the fuel's lower octane, higher flame speeds, lower stoichiometric air-fuel ratios, and increased densities. High flame speeds implied less spark advance, but the lower octane meant the knock-limited spark advance (KLSA) was also retarded, which meant that the engine needed less advance even at peak tune and that less spark advance was safe. Since the KLSA retard was greater, the Honda engine builders made a small reduction in compression ratio to allow optimal timing with sufficient knock margin. The net effect was to reduce thermal efficiency while increasing what Honda calls "combustion toughness," i.e., the ability to maintain good combustion under nonideal conditions of low load, lean mixtures, low temperature, etc. The higher fuel energy more than made up for the reduction in thermal efficiency, increasing power and responsiveness. Fuel consumption by weight increased due to the reduction in stoichiometric mixture, but fuel consumption by volume decreased due to the higher density of the fuel. This situation provided flexibility at critical circuits, enabling the engine to be calibrated closer to peak power air-fuel mixtures, maintaining engine output.

During the 1991 season, Shell implemented a program that could quickly optimize the main and relevant engine and fuel variables, identify potential qualifying fuels, test them, and explain the observed effects on performance. Using the program, and the high energy components, it was possible for the first time to design fuels optimized for qualifying (where only power mattered) and for racing (where power, fuel economy, and fuel weight were all important). Initial testing evaluated fuels on a test-bed engine at peak power rpm to measure torque at various mixtures and spark timing, which allowed building fueling and timing maps that could be used for qualifying. Fuels were also tested on a track under realistic conditions. Some fuels were evaluated for combustion characteristics in an engine equipped with a laminar flow meter in the intake, cylinder pressure transducer, and exhaust gas analysis equipment. The results correlated engine MBT to fuel flame speed, which could be related to knock margin in the more customary relating of knock margin to octane number alone, which is otherwise only valid for fuels with similar flame speeds. Where flame speed varies, it turned out to be necessary to consider not just MON, but flame speed index as well to estimate knock margin.

The above testing enabled Honda to successfully implement a mid-season engine design change, in which engine maximum speed was increased by increasing bore-to-stroke ratio. Increasing bore generally reduces combustion quality by increasing flame paths and reducing combustion chamber height, the high flame speed fuels enabled higher speeds with no sacrifice in combustion quality.

The most important engine change relating to the joint fuel-engine development program was the compression ratio reduction. Although combustion efficiency theoretically increases with increased in compression ratio, in practice efficiency peaks at 13:1 compression, due to the effects of dissociation and heat loss and the fact that, eventually, the large bores and valve notches required for raising compression are detrimental to combustion chamber geometry. Changes in the piston crown to lower compression in the Honda engine actually improved chamber geometry. By the end of the 1991 season, the Honda engine was making 90 more horsepower than at the first race, half of which came from the fuel-combustion system development.

A simulation program was developed to estimate the net effect of certain fuel changes, taking into account circumstances such as an increase in power, which might be offset by an increased fuel load, and the effect of various track lengths. Increased power is most evident on high-speed tracks; increased weight has the most effect on medium-speed tracks, where more time is spent braking and accelerating. As predicted, the Shell high-power fuels had little effect on short tracks but a great effect on high-speed tracks, which amounted to roughly a minute saved in the race—clearly enough to affect the outcome.

1992

In 1992, FIA required unleaded fuel and imposed density limits on fuels, the goal of which was to make F1 fuels more like commercial gasolines. All "power-boosting additives" (components not found in commercial gasolines) were banned. These included higher energy diolefins—olefins with two double bonds. Variations in the relative proportions of legal ingredients were still permitted.

The joint Shell-Honda-McLaren fuels-engine development program shows the delicate and interdependent nature of fueling very high-performance engines.

8

Alcohols

Both methyl alcohol, or methanol and ethyl alcohol, or ethanol have extensive advantages as high-performance fuels compared to gasoline. Methanol, written as CH_3OH, is usually synthesized from natural gas or coal; ethanol (C_2H_5OH) is fermented and distilled from corn or other biomass. Both are distinguished from hydrocarbon fuels by their hydroxy radical (OH), which produces alcohol's electrical polarity and affinity for water (H_2O, or H+ OH-). With stoichiometric specific energies (SE) of 3.08 and 3.0, respectively, both are slightly better than gasoline (SE of 2.92). The lower heating value of low-carbon alcohols (76,000 BTUs per gallon for ethanol, 57,000 BTUs for methanol) is more than offset by the favorable stoichiometric air-fuel ratio. At stoichiometry, methanol makes slightly more power than gasoline. However, methanol makes maximum power at a 4.0:1 air-fuel ratio, some 30 percent richer than the 6.45:1 stoichiometric air-fuel ratio, a condition under which the SE is vastly better than that of gasoline, which makes maximum power at only 20 percent richer than stoichiometric. Methanol's (and, to a degree, ethanol's) SE advantages as a racing fuel are magnified by the cooling effect of its high heat of vaporization, which significantly chills the inlet charge, providing improved engine volumetric efficiency. As a street fuel, methanol generally produces less HC pollution than gasoline. Its hygroscopic chemical properties can cause problems in alcohol-gasoline mixtures if enough water is present (as little as 0.1 percent can do it, depending on temperature and the aromatic content of gasoline) to cause phase separation, in which an alcohol-water mixture separates from the gasoline. Alcohol motor fuels, at a minimum, make similar power to gasoline; under the right circumstances, they make much more.

The combustion reaction of methanol is:

$2CH_3OH + 3O_2 \rightarrow 2CO_2 + 4H_2O$, which yields -22.7 kJ per gram.

History of Methanol

In the two centuries before WWI, methanol was produced from wood by destructive distillation, in which wood is heated in the absence of air, producing so-called wood alcohol. Essentially, methanol was distilled from a liquid produced during the manufacture of charcoal, producing three to six gallons of product from one ton of wood. By the mid-19th century, methanol was widely used in France as a fuel for cooking and heating. It was used for lighting until replaced by the more luminescent kerosene around 1880. In 1905, French Chemist Paul Sabatier invented a process, later implemented by the Badische Company, that synthesized methanol by combining carbon monoxide and hydrogen. The synthetic methanol was much less expensive than wood alcohol, bringing an end to a major wood alcohol industry. In the early 1900s, methanol was used as a motor fuel until forced off the market by low-cost gasoline.

In 1926, Commercial Solvents Corporation began producing methanol using a new process using hydrogen and carbon dioxide formed during the fermentation of corn. Later, Commercial Solvents began using coal as the source of the synthetic gas mixture. In the Haber process, coal was combusted to produce so-called synthesis gas hydrogen, carbon monoxide, and carbon dioxide, which could be converted to methanol by chromium-oxide and zinc-oxide catalysts under high temperature and pressure. Synthesis gas can be produced by the partial oxidation of any fuel containing carbon with oxygen or water. Because synthesis gas from coal has a sub-optimal hydrogen content, it must be treated during manufacturing. Methanol production from hydrogen-rich natural gas (which is mainly methane) usually makes use of excess hydrogen gas in an adjacent manufacturing process that produces ammonia. The availability of natural gas through the 20th century expansion of the petroleum industry has led to the replacement of coal as a methanol feedstock. Over 90 percent of methanol is now produced using natural gas, due to its high hydrogen content and the lower capital and operating costs of plants for natural gas methanol production.

In 1966, the single most important improvement in methanol production occurred when Imperial Chemical Industries in the U.K. developed a low-temperature and pressure-synthesizing process, using a copper-based catalyst at 500 degrees and 50 atmospheres. This lowered both capital and operating costs and eliminated the need to add CO_2 to balance the carbon-hydrogen ratio.

Today, most methanol is used as an ingredient in the manufacture of other chemical products, such as formaldehyde, acetic acid, acetic anhydride, and various other chemical solvents. By the mid-1980s U.S.

methanol production consisted of 1.2 billion gallons, the energy equivalent of roughly 16 million barrels of crude oil (about a half of one percent of the country's energy needs). Of this, perhaps 25 percent was used in the transportation sector, 95 percent of that in the manufacture of the gasoline additive MTBE (which has certain advantages over straight methanol), to oxygenate unleaded street gasoline in order to reduce air pollution, and used in racing gasoline to increase octane ratings and increase the fuel's specific energy.

Although it can be made from wood, lignite, or coal, or municipal, agricultural, or forestry waste, most methanol is now produced from natural gas. It turns out that methanol has advantages over natural gas for transport to markets from remote wellheads, which could make it increasingly feasible to build methanol production plants at the wellhead to make use of natural gas that would otherwise be wasted. Methanol can be transported using pipelines, like the Alyeska line in Alaska, that were designed for transporting petroleum, whereas natural gas requires construction of an entirely separate pipeline and removal of CO_2 and other higher hydrocarbons that will condense in pipelines. Methanol tankers and pipelines must be lined with zinc silicate or stainless steel and require fire detection and suppression systems.

With the increasing focus on the commercial production of energy from renewable resources, renewed production of methanol from biomass could make use of forestry and wood-processing residues, crop wastes, animal wastes, and crops specifically cultivated for energy production, or even secondary sources such as urban waste.

History of Ethanol

Ethanol has been around since ancient times as an ingredient of alcoholic drinks produced by fermentation. It has been made from virtually all agricultural products containing carbohydrates. Fruits, grains and vegetables with sugars, or other carbohydrates that can be converted into starch and the starch into sugars, can be used to make ethanol through the action of yeast on sugars, followed by distillation. Pure ("neat") ethanol is described in proof, with 100 proof spirit being 50 percent alcohol and the rest mainly water; 200 proof is pure ethanol.

Ethanol is rather expensive as a motor fuel. With government subsidies, U.S. farmers began producing ethanol, mostly from corn, in the 1970s following the energy shocks of 1973 and 1979. Ethanol was used as a 10 percent gasoline extender, marketed as "gasohol."

Following the international energy crises, the government of Brazil accelerated conversion of the country's motor transport fleet to nearly neat ethanol use in order to lower the consumption of imported petroleum. The number of vehicles using ethanol fuel increased from 235,000 to over 8 million between 1950 and 1980. By the mid-1980s, domestic Brazilian ethanol production amounted to 2.8 billion gallons of ethanol per year; nearly 500,000 new ethanol vehicles were being produced a year, and hundreds of thousand of older vehicles were converted from gasoline to run on ethanol. Initially, the use of gasoline was discouraged by halting sales of gasoline on weekends. Brazil's ethanol was produced from sugarcane, which already contained sugar and therefore eliminated the steps in production in which biomass is converted to starch, and starch into sugar. The Brazilian experience showed the viability of alcohol as a mass-produced motor fuel on a national level. Later on, when the price of sugar increased greatly, Brazil largely converted to a mixture of methanol and gasoline.

In the U.S., ethanol plants vary in size from small farm-based operations, which produce 40,000 gallons per year, to big commercial plants making millions of gallons a year. There have been many small personally owned ethanol fuel plants in the U.S.

Methanol and Air

Methanol dissociates into carbon monoxide and hydrogen as the temperature goes up during combustion, so the combustion properties of methanol are similar to those of carbon monoxide and hydrogen. Hydrogen's flame speed

	Methanol	Ethanol	Gasoline
Oxygen content, wt%	50.0	34.8	0
Boiling Point °C	65	78	35-210
Lower Heating Value, MJ/kg	19.9	26.8	approx 42.7
Heat of Vaporization, MJ/kg	1.17	0.93	approx 0.18
Stoichiometric air-fuel ratio	6.45:1	9.0:1	approx 14.6:1
Specific Energy, MJ/kg per air-fuel ratio	3.08	3.00	approx 2.92
RON	109	109	90-100
MON	89	90	80-90

Properties of ethanol and methanol as alternative fuels. Low-carbon alcohols are superior to gasoline as racing fuels. Disadvantages include alcohol's corrosive nature and the fact that neat alcohol burns with an invisible flame.

is considerably greater than that of petroleum fuels, and this is reflected in the flame speed of methanol, which is higher than iso-octane below a fuel-air equivalence ratio (ER) of 1.3, and especially in lean air-fuel ratios. Methanol has much wider misfire limits than gasoline, typically .2 ER units leaner than gasoline, allowing methanol engines to achieve high efficiency and clean emissions by running fairly lean.

Methanol Fuel Engines

Only in racing, where peak power production requires the highest specific energy from a fuel, is neat alcohol used. For street use, gasolines or other fuels are frequently blended with ethanol and methanol, usually in quantities from 10 to 50 percent. As the percentage of alcohol in the fuel mixture increases, the air-fuel mixture must be progressively richened. Modern oxygen-sensor-equipped, closed-loop electronic carb or fuel injection systems will automatically calibrate the air-fuel mixture to achieve target levels of exhaust gas oxygen in fuels with low percentages of oxygenated fuels, which include all ethers and alcohols.

Brazil began developing flexible vehicle fueling systems in the 1970s and 1980s, which could use anything from straight gasoline to 50 percent alcohol fuels by jetting carbs for the rich air-fuel mixture required by maximum alcohol content, and then electronically leaning the mixture for higher gasoline percentages with electronically controlled air bleeds, according to the quality of combustion. In the 1980s and 1990s, U.S. vehicle manufacturers began to experiment with electronically injected flexible fuel vehicles, which measured the alcohol content of fuel on the fly, according to its specific gravity. These modern flex-fuel vehicles handle anything from straight gasoline to neat alcohol.

Alcohols have excellent octane ratings (109 RON, 89–90 MON, or about 10 points higher than street gasoline) with high sensitivity (RON-MON spread), which allows more efficient high compression ratios. Since alcohol fuels chill intake air significantly, the CFR octane test on alcohols requires special test equipment to reheat intake air, especially for the MON test, a procedure which nullifies alcohol's charge-density advantages and also produces octane ratings that are probably lower than those of alcohol running in a real situation where the cold intake charge would tend to lower peak combustion temperatures and reduce the propensity of an engine to knock. Since engine efficiency improves 16 percent as compression ratios increase from 8:1 to 18:1, alcohol's octane rating translates directly into usable power on engines designed specifically for alcohol fuels.

The high heat of vaporization of alcohols improves engine volumetric efficiency, and their flame speed is high; they burn cleanly; and their rich flammability limits allow high power-producing rich mixtures. Alcohols produce higher volumes of combustion products than gasoline, making higher cylinder pressure. They also burn with lower flame temperatures and luminosity, so less heat is lost into an alcohol engine's cooling system.

	Methanol optimized	Ethanol optimized	Gasoline	
Performance				
0-100 km/h (sec)	9.0	10.1*	10.0	10.5*
V_{max} (km/h)	214	207*	210	205*
Fuel consumption				
City cycle (l/100 km)	27.0	22.8*	17.3	16.8*
90 km/h	15.8	13.9*	9.1	9.4*
120 km/h	18.9	17.0	11.3	11.7*
*Automatic transmission version				

Performance comparision of alcohols with gasoline.

(But this also means that deadly fires can be burning in the case of an accident, with invisible transparent flame!)

Due to the lower heat content of alcohols, volumetric fuel economy is always lower than with gasoline, requiring larger fuel tanks to go an equivalent distance. It requires about 1.8 times as much neat methanol to provide range equal to gasoline.

Neat alcohols' low vapor pressures (4.6 psi for methanol, and 2.3 for ethanol vs 8 to 15 psi for gasoline) produces hard starting in cold weather. The minimum ambient temperature for starting neat methanol engines without starting aids is between 32 and 60 degrees, and for neat ethanol, 110 degrees. In Brazil, ethanol vehicles have used auxiliary gasoline-ethanol mixtures for starting, and propane has been used to help start methanol vehicles. Computer-controlled vehicles with alcohol-starting logic easily start with near-neat mixtures of alcohol and volatile primers such as gasoline or dimethyl ether. Primers may have side effects such as water sensitivity, hot-weather drivability problems, and excess evaporation losses.

Ignition timing can often be advanced when converting an engine

Dee Howard's custom '32 Pierce-Arrow speed record car is designed to exceed 200 mph on alcohol.

from gasoline to alcohol because of the increased octane rating.

Fuel mileage increases as the alcohol content of gasoline-alcohol fuels approaches 10 percent, then begins to decrease again, as mixtures become increasingly rich. A test vehicle decreased from about 18 mpg with 10 percent ethanol to 13 mpg with 50 percent ethanol. The same vehicle got 14 mpg with 30 percent methanol.

Although Brazil has millions of vehicles running ethanol fuels, that country has no government-mandated emissions standards, so there is little knowledge regarding the effect of alcohol fuels on the environment. However, that situation is turning around as increasing numbers of alcohol vehicles are tested in the U.S. It is known that evaporative emissions from fuels tanks and carbs increase slightly with higher percentages of alcohol in fuels. For gasoline-type engines running on alcohol, lower peak combustion temperatures, and the ability to run lean mixtures without misfires, tend to reduce NO_x emissions. However, with optimized engines running higher compression ratios, NO_x emissions increase. Mixtures of 30 percent ethanol or methanol reportedly have produced NO_x emissions twice that of gasoline. With additional emissions control equipment, NO_x emissions are similar to those of straight-gasoline-fueled vehicles. Methanol and methanol-gasoline blends had clear HC emissions advantages, and CO emissions are unchanged or improved with alcohol fuels in optimized engines.

Methanol exhaust gases have a lower photochemical reactivity than gasoline exhaust gases in terms of smog formation. Unburned fuel and formaldehyde formation are problems compared to gasoline, although three-way catalysts, EGR, and closed-loop engine controls can reduce this, however, methanol's lower exhaust gas temperature reduced the efficiency of the catalyst.

Methanol and ethanol are corrosive to galvanized steel, aluminum, copper, brass, magnesium, die-cast zinc, and Terne metal—the rust-resistant lining material in many fuel tanks. Alcohols can also attack plastics, fluorocarbons, and elastomers, including carb floats, which can swell and stick. Fiber gaskets, fuel hoses, and pump diaphragms have all been known to swell or harden and crack. Distribution and storage must take the above into account, and tanks should take into account that the vapor above methanol is likely to be flammable between 45 and 110 degrees F, though not when gasolines or other hydrocarbons are added. Neat methanol burns with an almost invisible flame, but relatively small amounts of gasoline will add to the luminescence of the flame.

Neat methanol and ethanol have been known to cause accelerated cylinder and ring wear, possibly due to cylinder wall washing by rich mixtures during warmup. Some experts speculate that methanol fuels may be creating formic and performic acids during combustion; these are corrosive to iron. Chrome rings reduce the problem. Many experts recommend oils specially formulated for alcohol engines to prevent wear due to alcohol attacking conventional oils.

Methanol as Hazardous Material

Methanol ("wood alcohol") is deadly to humans. Although it has an initial transient narcotic inebriating effect, in 1 to 72 hours, toxic symptoms begin to appear, including weakness, dizziness, headache, sensation of heat, nausea, abdominal pain, followed by vomiting, dyspnea, acidosis, visual disturbances, convulsions, coma, and death. Methanol damages the central nervous system, particularly the optic nerve, producing photosensitivity and temporary or permanent blindness. Ingesting 50–100 milliliters is considered a fatal dose, and as little as 25–50 milliliters has often been fatal if not treated immediately. Siphoning by mouth should be strictly avoided.

Inhaling high concentrations of methanol can cause poisoning from brief exposure. One-thousand parts per million (ppm) in air will irritate the eyes and mucous membranes. The odor of methanol is barely detectable at 2,000 ppm; 5,000 ppm can cause stupor or sleepiness. One to two hours exposure to 50,000 ppm results in deep narcosis and possibly death. Two-hundred ppm is considered the safe upper limit for constant exposure for eight

Alcohol fuel system in the Pierce-Arrow trunk shows stainless steel fuel system required for corrosive methanol, as well as the fuel supply hardware.

hours a day in a 40-hour week. Methanol can be absorbed through the skin.

People exposed to methanol should get immediate first aid and then the attention of a doctor. A patient should immediately be removed from any contact with methanol vapors and should get artificial respiration for stopped breathing. Methanol-contaminated clothing should be removed and the skin washed. Contaminated eyes should be washed with water for 15 minutes while being kept open, and the patient taken to an ophthalmologist immediately. Patients who have ingested methanol should have the stomach emptied as soon as possible by induced vomiting.

Methanol as a Racing Fuel

Methanol is well established as a racing fuel, particularly in Indy and drag racing, and to a lesser extent in circle track racing, either neat or in lower blend percentages. Neat ethanol, intermediate in properties between methanol and gasoline, has also been used successfully in racing, although it is more expensive and is not as pronounced in its advantages over gasoline compared to methanol.

Methanol makes more power than gasoline—a mid-range torque increase of 8–12 percent on unoptimized engines is typical—because a given amount of air sucked into an engine can burn much more methanol at stoichiometric mixtures than gasoline; so much more that the total specific energy of alcohol slightly exceeds gasoline. What's more, maximum power on methanol is achieved at mixtures very rich of stoichiometric, and at rich mixtures the specific energy of methanol is much higher than gasoline. Methanol's high antiknock rating permits very high compression ratios for high efficiency and power, typically as high as 16:1 or more. It also permits extremely high boost in turbocharged engines. The high heat of vaporization chills intake air enough to significantly improve engine VE and, so, power. Methanol produces much less severe engine mechanical stresses than nitromethane while still permitting relatively high power, making it suitable for long-distance racing, such as the Indianapolis 500; other circle track racing, and specialty events such as tractor pulls. Methanol is easily available and relatively cheap.

However, methanol's low energy content means that fuel consumption is a problem when

Ancient Pierce-Arrow V-12 engine has been updated by River City Products to include forged pistons, improved induction, and twin turbochargers. Mechanical methanol injection provides improved power and improved combustion temperatures compared to gasoline fueling.

Fueling Engineering's high output Olds Quad 4 was reconfigured to achieve 1,300-plus horsepower on methanol with twin turbos and a supercharger! Like most engines running neat methanol, the Fueling version engine is very cold-blooded and runs terribly until reaching operating temperature (which is why 15 percent gasoline is often blended with methanol).

operating at the rich mixtures required for maximum power, so methanol racers need a large fuel supply to have much range. Methanol engines run quite cool at rich maximum-power air-fuel mixtures; at leaner mixtures (and alcohol lean burn limits allow quite lean mixtures without misfiring), the engine will run hotter with reduced power. Cold starting with neat methanol can be difficult in cool weather due to the low vapor pressure, and typically 15 percent gasoline or ether is added to overcome this problem. Unlike gasoline—a blend of components with a broad range of boiling points—methanol fuel vaporizes all at once at a single temperature. This can cause carburetion and drivability problems, which is another reason why it is common to blend in some gasoline or other hydrocarbon fuel. Port fuel injection is effective in preventing or reducing these problems. Methanol and other alcohols do not respond well to lead octane improvers.

Ethanol and the higher carbon alcohols butanol and propanol are increasingly similar in combustion characteristics to gasoline as the percentage by weight of oxygen in the fuel goes down. Ethanol's maximum power in racing occurs at an air-fuel ratio of 7:1. Ethanol and higher alcohols are much more expensive to produce than methanol.

Brake-specific fuel consumption (BSFC) is the fuel flow, in pounds per hour, per horsepower. For methanol it is typically 1.0–1.2, but can be as good as .88 or as bad as 1.35, compared to gasoline's BSFC of somewhere around .4–.6. Carburetors and injection systems designed for methanol require modifications to handle fuel delivery requirements 2–2.5 times higher and to prevent methanol from attacking vulnerable metals, plastics, and rubbers. The main wells and metering body cross-sections of Holley carbs are opened up as much as possible for methanol use. Supplemental passages are drilled from the float bowls to each main well, just above the main jet, enabling greater fuel flow without requiring externally larger main jets. Needles and seats are steel with the largest possible opening, and Holley has recommended relatively high fuel pressures to the float chamber. Brass floats initially replaced Nitrophyl, which will not hold up in methanol. Later on, Holley began supplying hollow plastic floats. Large accelerator pumps are used. Because castings are not plated, and the pump diaphragm is unchanged, racers routinely drain methanol carbs and frequently tear down and clean and change the diaphragm and gaskets.

As a slower-burning fuel, methanol requires (and can tolerate) more ignition advance than gasoline. In order to make high peak horsepower on methanol, it is vital to optimize the engine with very good ignition to get the flame started fast, due to the fact that alcohol is a slow-burning fuel. Extremely high rpm engines that are not optimized for methanol will make less power than on gasoline because the flame speed may be too slow for complete combustion in time. A high-domed piston used to achieve high compression can slow flame travel enough at high speed such that power can actually be less than with gasoline.

Methanol not only cools the intake charge, but the combustion temperatures are typically 100–150 degrees F lower than with gasoline. Engines with good, efficient cooling systems and good combustion efficiency can safely operate at the lower end BSFC, but if combustion or cooling efficiency is suboptimal, more methanol will be needed to keep the engine running at a reasonable temperature, which will hurt torque and power and push BSFC above 1.15. To accommodate over twice the fuel flow, a 50 percent increase in the diameter of fuel plumbing, jets, injectors, etc., is required.

Because methanol mixtures are so rich compared to gasoline, methanol vapors will actually displace enough air in the intake charge that compensation is required in the form of bigger intake runners to prevent loss of air-flow. Enlarging the runners should be approached conservatively to prevent excessively low intake gas velocities, which hurt VE and enable fuel to fall out of suspension and puddle in the intake manifold. For similar reasons, straight, single-plane manifolds should be used; where turns are unavoidable, they should be large in radius. Methanol should really be port-injected where rules permit. "Wet" manifolds for alcohol need to operate at higher temperatures to aid in good distribution, to minimize fuel drop-out and puddling, and to maintain combustion efficiency by vaporizing as much of the fuel as possible before it enters the combustion chamber, which requires nine times more heat than an equivalent gasoline engine.

Where ambient temperatures are below 50 degrees F, heating the intake manifold plenum will improve engine performance throughout the entire rpm range, particularly low-speed drivability and mid-range torque and response. Running coolant through a methanol-cooled intake will remove some of the cooling load from the radiator and cooling system. A circle track V-8 running methanol will do best with coolant at 190–120 degrees F.

To compensate for slow burning at part throttle load below 3,000 rpm, some expert tuners decrease centrifugal (rpm) advance so they can increase the amount of initial advance at idle. Tuners have been known to provide no rpm advance, to get a smooth idle, crisp low speed response, and drivability. Smaller engines and longer cams need more initial advance.

Some engine builders have increased rod lengths in relation to stroke to extend the torque advantages of methanol. Where a methanol engine uses longer rods and large intake runners, cam selection is particularly important—it is easy to "overcam" an alcohol engine. Increased duration, along with large ports and long rods, can allow the intake gas velocity to drop below a critical level and cause rapid VE drop-off.

Nitro, Monopropellants, and Rocket Fuels

Many explosives and ultra-high-performance fuels contain nitrogen compounds and make use of the fact that these compounds are relatively unstable and yield tremendous energy when they break down. Examples include explosives like nitroglycerin, which is so unstable a sudden impact can cause it to explode.

Other well-known destructive nitrogen compounds include trinitrotoluene (TNT), and nitrocellulose (gunpowder), as well as rocket fuels like nitromethane and hydrazine, which have also been used as ultra-high-performance automotive fuels. These explosives and fuels contain oxygen in the nitro group (NO_2) and can break down to gaseous products of large volume and heat without further oxygen! The hot, rapidly expanding gases produce a violent pressure surge and damaging shock wave of explosive force. Nitro compounds have a very favorable products-to-reactants ratio. For example, two moles of TNT produce 20 moles of gaseous product, plus energy:

$$2C_7H_5N_3O_6 \rightarrow 12CO + 5H_2 + 3N_2 + 2C + energy$$

Saturated hydrocarbons like methane can undergo substitution reactions, in which one or more of the hydrogen atoms is replaced by a halogen, nitro, or hydroxyl group. Nitromethane is a substituted methane hydrocarbon. So is methanol, but where nitromethane has a NO_2 nitro group in place of a hydrogen atom, methanol has a polar hydroxyl group. Due to the presence of oxygen in the fuel, less oxygen is required per unit weight of fuel, meaning a given amount of air drawn into an engine can burn more fuel.

Utilizing nitromethane to improve the power of a piston engine involves special practices and techniques to produce reliable power. Specially designed blocks, heads, pistons, connecting rods, and crankshafts are needed for nitro and specialty fuel engines. The high moles of product gases versus reactants results in very high cylinder pressure, which translates to horsepower. The high flame speed of nitromethane combustion makes it very desirable as a fuel in dragsters and funny cars, the high rpm of which (8,500–10,000) leaves little time for combustion. Nitro's fast burning rate reduces time-related power losses during the Otto engine cycles, yet fuel cars typically expel burning nitro directly out of the exhaust, because of the massively rich mixtures required to combat detonation.

Nitroparaffins

Nitromethane, the nitroparaffin used most commonly as a racing fuel, and the main focus of this discussion, can be combusted in reciprocating spark-ignition race engines at incredibly rich mixtures. In fact, it can even be combusted as a monopropellant, with no air at all. The chemical formula for nitromethane is CH_3NO_2. Other nitroparaffins include nitroethane ($C_2H_5NO_2$) and nitropropane ($C_3H_7NO_2$) and have somewhat similar, but less extreme properties.

History of Nitroparaffins

Nonhydrocarbon fuels were widely used during the 1930s in European Grand Prix cars. German Mercedes-Benz and Auto Union racers, and Italian Masaratis and Alfa-Romeos used ethanol/

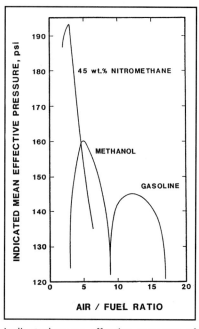

Indicated mean effective pressure of nitromethane, methanol, and gasoline at various air-fuel mixtures. *SAE*

methanol-based fuel blends with acetone, ether, benzene, nitropropane, and nitrobenzene components added for increased power. The ethanol was increased when fuel economy was important. A blend of 86 percent methanol, 8.8 percent acetone, 4.4 percent nitrobenzene, and 0.8 percent ether could actually be obtained by special order from the Standard Oil Company of New Jersey.

Nitromethane was first used as a liquid rocket fuel in the 1940s. Following WWII, automobile racers tried nitromethane as a primary automotive fuel cut with methanol in flat-head Ford V-8-powered drag cars. Nitromethane was simultaneously being tried in fractional horsepower model airplane

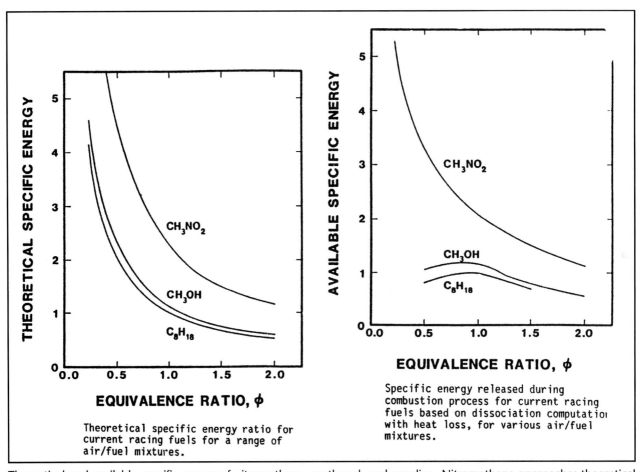

Theoretical and available specific energy of nitromethane, methanol, and gasoline. Nitromethane approaches theoretical limits due to its greater rich flammability limits. Nitro can actually combust without any air, as a monopropellant, and has been used as a rocket fuel. *SAE*

engines, and many model car and airplane fuel blends available in hobby stores today are still mostly nitromethane.

The first racing engines burning nitro were naturally aspirated, carburetted production engines, but Top Fuel engine development has produced powerplants made to survive 5,000+ horsepower drag runs at full power on nearly pure nitromethane. These engines are typically supercharged to more than 6 atmospheres, and run at 8,000–10,000rpm. In 1995, a Top Fuel dragster weighing less than 2,000 pounds could accelerate to over 300 mph in less than five seconds within a quarter of a mile. Nitro-burning engines—both supercharged and naturally aspirated—have also been used in drag boats, pulling tractors, unlimited class air race aircraft, and in some Indy cars.

Few scientific studies have been conducted to evaluate the effect of nitromethane on the power output of internal combustion engines. During the 1950s, some tests were made under laboratory conditions using low percentages of nitromethane in methanol, but most testing took place under drag-race conditions. The conditions of neat or near-neat nitromethane combustion are so harsh that an engine can hardly last long enough for meaningful dyno testing, so horsepower output is based on estimates derived from weight and acceleration of a drag vehicle.

Extremely rich nitro mixtures are used in drag engines to control combustion temperatures and knock, and to help cool the engine. Top Fuel motors require prodigious amounts of fuel to produce peak power in blown 500-cubic-inch motors—over 7,000 pounds per hour was typical when Top Fuel motors were making half the horsepower they are calculated to produce today. This implies fuel usage in the 100–200 pounds per minute range, which is something like a gallon of nitro every three seconds—faster than you could dump it out of a bucket! These excessively rich mixtures result in incomplete combustion and fuel still in flames as it leaves the exhaust pipes.

Nitromethane combustion under Top Fuel conditions results in extremely fast and erratic pressure rise. The high pressure and temperatures can lead to pre-ignition, which adds to the instability of nitromethane combustion and can lead to dieseling when the ignition is shut down. Nitromethane ignites in air at 785 degrees, a temperature easily achievable by the heat of compression in a high-compression piston engine, and which is thought to decrease with higher pressures. With the huge amounts of liquid fuel being dumped into a Top Fuel motor, much of the fuel may exist in liquid globules, which can accumulate and then combust unpredictably by pre-ignition. This can cause severe engine damage, including piston "backsiding." Toluene and benzene have been used as additives to nitromethane to help control pre-ignition. As a potential monopropellant, nitromethane has flammability limits vastly greater than gasoline. Lean nitromethane mixtures can result in disaster, since the flame speed slows markedly in an oxygen-rich environment, and flashback explosions can result when the intake valve opens on long-overlap engines.

Nitromethane is corrosive to the aluminum and magnesium heads and blocks forming the combustion chamber of a racing engine, forcing racers to drain and purge the fuel system after racing to avoid damage.

Nitromethane Combustion Theory

The fact that Nitromethane can burn without air makes it both useful as a fuel and dangerous. Nitro has a low heating value but a specific energy over twice as great as gasoline: It is possible to dump nitro into an engine in air-fuel ratios as high as 1:1. A naturally aspirated engine burning nitro instead of gasoline has the potential to make over double the power. Nitro, however, is a harsh fuel, very prone to knock, with the potential to produce severe engine stress and damage, even in highly built race motors. It is used as a primary fuel only in drag racing, where engines only run a few seconds at full power, and even then it wears out these engines quickly—a Top Fuel engine that has run 60 seconds is an old timer. Nitro is sometimes used for longer duration circle track racing as a power-booster additive for alcohols or gasoline. Any motor burning a nitro blend must be built to stand up to severe stress and must have fuel systems designed to deliver air-fuel ratios significantly richer than air-gasoline mixtures alone.

Chemistry of Nitromethane

The reaction of nitromethane as a monopropellant is expressed as:

$$CH_3NO_2 \rightarrow 0.25CO_2 + 0.75CO + 0.75H_2O + 0.75H_2 + 0.5N_2$$

Nitromethane's stoichiometric combustion in air is expressed as:

$$CH_3NO_2 + 1.5O_2 + 2.82N_2 \rightarrow CO_2 + 1.5H_2O + 2.82N_2$$

Nitromethane's specific energy—the amount of energy delivered to the combustion chamber per unit mass of air—is much higher than that of gasoline, although the heating value per weight of fuel is low. It doesn't need much air, which is its main advantage—some would say its only advantage. The performance of any fuel in an engine relates to the release of combustion energy, which is modified by dissociation of the products of combustion and heat transfer into the combustion chamber. With large products-to-reactants ratios, high heat of vaporization and extremely high specific energy, nitro is a formidable fuel, and would be more formidable if not for its problems.

Stoichiometric combustion of nitromethane yields 5.82 moles of product for 4.57 moles of intake charge reactants, a favorable ratio of 127 percent compared to methanol and gasoline's 106 percent ratio. Dissociation—in which molecules split into simpler molecules or individual atoms under high temperatures and pressures—increases nitromethane's products-to-reactants ratio to 130 percent, and with very rich mixtures (half stoichiometric) increase it to 160 percent, partially due to the appearance of carbon monoxide.

Factors Affecting Piston-Engine Nitromethane Combustion

Nitromethane is a harsh fuel, with a strong tendency to knock, so nitro engines typically run very low compression ratios. The combination of nitromethane's flame speed, the high rpm of fueler drag motors, and the requirement to set ignition timing at a point that minimizes detonation typically results in Top Fuel drag engines that expel large amounts of still-burning nitro during the exhaust stroke. Experiments by Bob Norwood and C.J. Batten have indicated the value of multiple-spark plugs (up to three per cylinder) in increasing the efficiency of nitromethane combustion by lighting the air-fuel mixture in multiple places, which causes more burning during the combustion stroke and increases power considerably. Since racers typically run extremely rich air-nitromethane mixtures to control knock, any method of improving combustion while controlling knock has the potential to increase power.

In 1985 Bush, Germane and Hess tested engines with nitromethane fuel combined with methanol and various higher-carbon nitroparaffinic co-solvents. These tests, at BYU, investigated various engine operating parameters on power output and combustion severity, including compression ratio, ignition timing, air-fuel mixture, and co-solvent effects.

Testing a 5:1 compression engine with 30 degrees spark advance at slightly over 1,000rpm at stoichiometric mixtures showed that increasing nitromethane fuel from zero to 50 percent in methanol increased power by roughly 45 percent, while knock increased roughly 20 percent. Using a 40 percent nitromethane mixture, both maximum power and maximum knock occurred at 20 degrees BTDC and dropped quickly with increased or

Nitromethane is only suitable as a neat fuel for short-duration drag racing, but cut with alcohol, which tempers the nitro's combustion and high tendency to knock, the fuel has been used in Indy car racing.

decreased spark advance. Since knock fell off more rapidly than power on the retarded side of peak power, Bush et al recommended setting spark advance slightly retarded of peak power to control knock while maximizing power.

Increasing compression ratio from 5:1 toward 8:1 increased both power and knock, but power increased faster. Changing mixture strength revealed that power increases steadily while knock decreases rich of a 1.1 equivalence ratio all the way to 1.7 and beyond. In the final testing, when replacing the methanol co-solvent with a nitroethane or nitropropane fuel, power increased with increased percentages of nitromethane (equivalence ratio was constant, so the air-fuel ratio became more fuel rich as nitromethane percentage increased), and knock severity decreased with increased nitromethane. This is opposite to the observed results with the methanol co-solvent, in which knock increased. It required 50 percent more nitromethane in methanol to achieve power levels equal to nitromethane and nitroethane, and at that level, knock was 50 percent greater than with the pure nitroparaffin blend. The experimenters speculated that this effect was due to a change in composition or temperature of the end gas, since no other engine conditions were changed. Nitropropane appeared to be a more effective co-solvent than nitroethane. The researchers tried adding gasoline-type TEL and MMT antiknock additives but found them ineffective at reducing knock, while hurting power slightly.

It is interesting to consider the effect of gasoline as a co-solvent in these tests, since the decreased oxygen content of nitropropane compared to nitroethane seemed to produce benefits...

Hydrazine and Other Nitrogen-Based Rocket Fuels

Hydrazine, with the chemical formula N_2H_4 (H_2-N-N-H_2) is a colorless, fuming, corrosive hygroscopic liquid with an ammoniacal odor which freezes at 2 degrees C and boils at 113.5 degrees. (Hygroscopic means it absorbs water easily, including water vapor from the atmosphere.) When mixed with nitromethane, it forms explosive salts (hydrazinium salt of methazanic acid) that require only the oxygen in nitromethane to combust.

The combustion of ordinary hydrazine is as follows:

$$N_2H_4 + O_2 \rightarrow N_2 + 2H_2O$$

History of Hydrazine

Hydrazine has application in rocket and jets fuels, and was first used as an automotive fuel additive in the late 1940s as a nitromethane additive. It has the potential to produce dramatic increases in power due to its high reaction rate, which produces extremely high cylinder pressures in a piston engine. And very little is required. Hydrazine has killed racers and spectators by literally blowing the heads off a racing engine, possibly due to the formation of explosive hydrazinium salts—a Class A explosive. Hydrazine continues to receive intermittent attention as a racing fuel power booster due to its potential to make incredible amounts of horsepower, even if engine longevity has been marginal. Naturally, it is illegal in most classes of racing.

Hydrazine has been blended with nitromethane, methanol, and gasoline for use in conventional piston engines by a few brave souls, but clearly its full potential for automotive racing has hardly been explored. It has also received some testing in special experimental reciprocating aviation engines as a monopropellant due to the high energy content and easy conversion to large amounts of hot gases.

Natural Gas and Propane

Natural gas is mostly methane, which is the simplest hydrocarbon and has the chemical formula CH_4. As recovered at the well-head, natural gas may also contain up to 20 percent higher hydrocarbons such as ethane (C_2H_6), propane (C_3H_8), and butane (C_4H_{10}), although these may be removed and sold as liquified petroleum (LP) gas, or "propane." When present in natural gas, these light hydrocarbons have similar properties to methane, however, with increasingly higher boiling points and decreasing octane

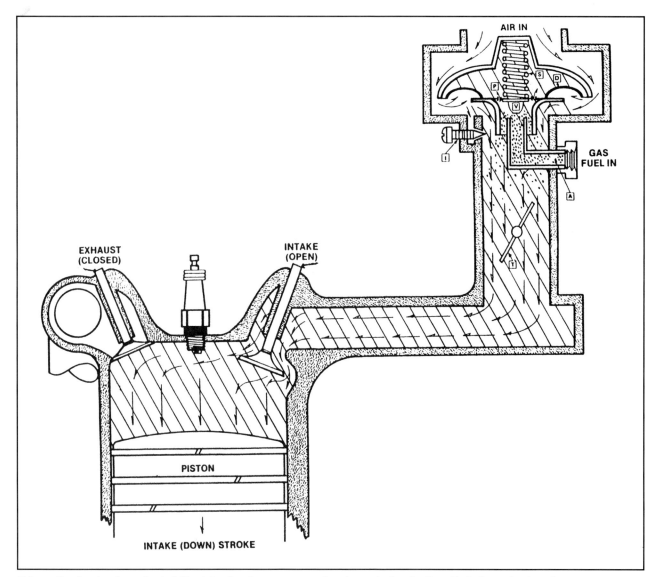

Schematic of natural gas "carb." The job of such a propane mixer is much simpler than that of a carb, since there are no vaporization issues to be handled. Separate circuits for idle, mid-range, and wide-open throttle are not required. *Impco Carburetion, Inc.*

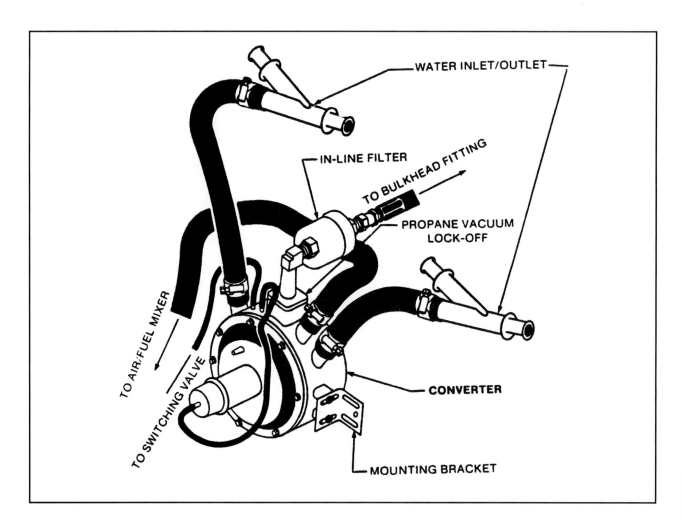

Propane converter supplies the fuel at a fixed pressure, regardless of the ambient tank pressure. Hot engine coolant keeps the regulator from freezing. *Impco Carburetion, Inc.*

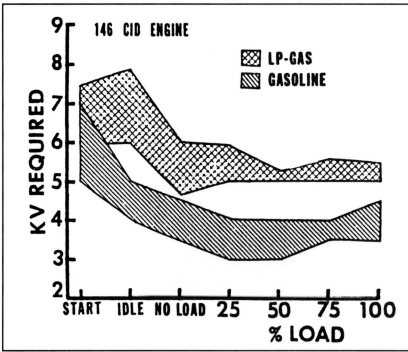

Propane secondary ignition voltage requirements are significantly higher than those of a gaoline-fuel engine. *Champion Spark Plug Company of Canada, Ltd.*

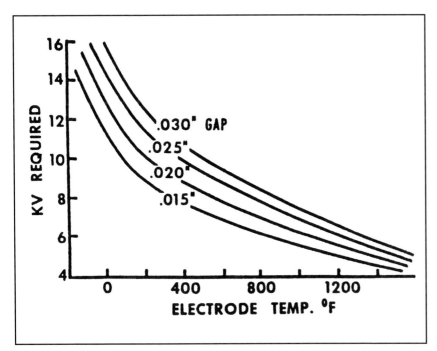

The effect of low heat range plugs on spark voltage requirements. Hot plugs require much less voltage to fire. *Champion Spark Plug Company of Canada, Ltd.*

ratings. As the number of carbon atoms increases, the carbon-to-hydrogen ratio increases, as does the molecular weight. Natural gas has the highest possible hydrogen-to-carbon ratio of any hydrocarbon, with one carbon atom tightly covalently bonded to four hydrogen atoms. It does not dissociate easily and therefore has extremely high octane (roughly 130), which qualifies it for consideration as a high-performance motor fuel. Flame speed is rather slow. Natural gas combusts very cleanly and produces very little oil contamination. It boils at extremely low temperatures, vaporizing at -260 degrees.

Natural gas has historically cost 1/2 to 1/3 the price of gasoline. Compared to gasoline, natural gas is dirt cheap—even when you consider that it contains less energy. One gallon of gasoline is equal to about 100 cubic feet of natural gas. Compressed natural gas contains 32,154 BTUs per gallon at 3,000 psi; liquified natural gas contains 83,000 BTUs per gallon. A power loss of 10–15 percent has been experienced in some gasoline-to-natural-gas conversions, although an engine optimized for natural gas could minimize or eliminate this disparity.

Propane

The chemical symbol for propane is C_3H_8. Propane is a straight-line saturated paraffinic hydrocarbon with a high ratio of hydrogen to carbon. Its excellent octane rating of 112 is 20 points higher than gasoline's, which permits efficient high compression ratios. That, coupled with propane's clean combustion and low oil contamination, is reason to consider it a "high-performance" fuel, even though it has a lower energy content than gasoline and requires a lot of air to burn, which means the specific energy of propane is relatively low. Propane contains 91,500 BTUs per gallon; gasoline ranges from 113,000 to 147,000 BTU per gallon. Propane boils at -42 degrees F. It generally sells for anywhere from half the price of gasoline per gallon to slightly more.

History and Production of Propane and Natural Gas

Historically, natural gas was wasted by flaring or burning-off at the well-head. The majority of propane and LP gas in the U.S. comes from natural gas wells and processing plants and is known as wet gas or well-head gas. Another one third of U.S. propane is a by-product of gasoline refining. Natural gas can also be synthesized by coal gasification, or from kelp and biomass. The supply of "wet" propane and natural gas has increased as petroleum producers have quit "flaring" and have begun to capture the gas. In the past, Saudi Arabia has flared off as much as 4 billion cubic feet of natural gas per day from enormous desert oil wells.

In the U.S., a large distribution network exists to supply propane for home heating, industrial uses, and for recreational vehicle heating and cooking requirements. Similarly, a substantial natural gas pipeline infrastructure exists in

Natural Gas Air-Fuel Ratios and CO Levels							
Natural Gas	-	18.0	*16.2*	14.8	14.3	13.8	-
Butane Propane	18.0	16.3	*15.4*	14.2	13.6	12.8	12.0
Gasoline	16.0	15.0	14.0	*13.0*	12.0	11.0	10.0
Carbon Monoxide %		0	1	2	3 4	5 6	-

Air-fuel ratios have a huge impact on carbon monoxide emissions. *Reston Publishing Co.*

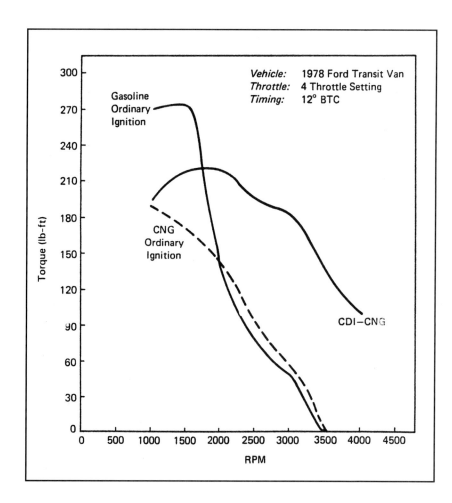

Engine torque with CD versus points ignition. At speeds of 1,700-plus rpm, natural gas with CD ignition produced significantly more torque than the same engine fueled by gasoline with points ignition. *New Zealand Energy Research and Development Committee*

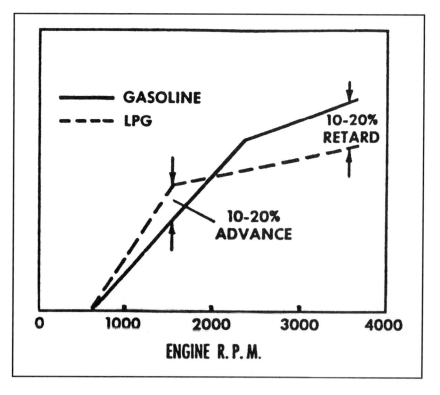

Propane distributor advance requirements. *Champion Spark Plug Company of Canada, Ltd.*

urban areas and some rural areas. By 1995, "Alternative Fuels—Propane" signs had begun to appear along interstate highways, indicating that facilities were available to refill large propane tanks in a timely manner, and the sellers were equipped to collect taxes on propane as a motor fuel.

As Brazil has led the world in alcohol-fueled vehicles, Italy has led in natural gas. With no native energy resources, Italy built a large-capacity pipeline to transport natural gas from Libya directly across the Mediterranean Sea, and for many years Italy has led the world in natural gas-fueled vehicles. By the early 1980s, Italy had an infrastructure of hundreds of liquid and compressed natural gas filling stations for motor vehicles.

In the 1990s, air pollution was the incentive for establishing natural gas as an important fuel in the U.S. Hundreds of thousands of vehicles have been converted from gasoline to straight propane or dual fuel gasoline/propane operation in the U.S. and Canada since 1980, in some cases with

Norwood Autocraft-designed Cadillac Seville is powered by clean-burning propane.

government subsidies. More recently, the federal government and state governments in Texas and California have begun to subsidize municipal and commercial organizations, as well as individuals, to convert vehicles to "clean-burning" natural gas fuel. At the same time, car makers haltingly have begun to supply factory original-equipment propane vehicles, with Ford being the first to offer these to the public.

Propane Combustion Theory

Unlike gasoline, lean mixtures of propane burn cooler. Rich propane mixtures, on the other hand, can burn up an engine quickly. Propane is generally converted from liquid to vapor in a vaporizer and delivered to a vapor-air mixer. Because it enters the engine cylinders in gaseous form, propane provides essentially no cooling effect on intake air, though there has been some recent experimentation with electronic injection of liquid propane.

Since propane is a gas above -40 degrees, distribution and atomization are not a problem, and no manifold heating is necessary or desirable for good combustion. With propane or natural gas fuel, there is no possibility of cold intake manifolds causing fuel condensation, since propane can normally only be liquified under pressure. Cold intake air will always improve VE, and hence performance, with any fuel—each 10 degrees of intake heating causes a one percent power loss, unless heating is required to produce a well-vaporized intake charge. All coolant- or exhaust-heated passages through the intake manifold should be blocked on propane-fueled engines. Engine coolant thermostats above 180 degrees should never be used with propane; 165-degree thermostats are best.

Like natural gas, propane provides essentially no lubricative effect on exhaust valves, and it additionally provides no means to get additives into the motor; exhaust valve recession can be a problem. It is a hot dry fuel that is hard on valves, easy on rings.

Correct air-fuel ratios are critical on propane- and natural gas-fueled vehicles. Rich mixtures typically increase exhaust gas temperature (EGT), head temperature, and valve face temperatures at least 40–60 degrees. Propane is not a fast-burning fuel; lean propane air-fuel ratios can easily slow combustion to the point where excessive burning continues after the exhaust valve is open, increasing EGT and decreasing valve life.

Engine Design and Tuning for Propane

Engine compression ratios should be raised to take advantage of propane and natural gas' high octane ratings. Propane fuel should easily support compression at least 2 to 3 points higher than an equivalent gasoline-fueled engine. Since propane is a cleaner-burning fuel than gasoline, in most cases exhaust emissions have been met even with increased compression (which tends to increase NO_x emissions due to increased combustion temperatures). At 15.5:1, propane's stoichiometric mixture is leaner than that of gasoline, but since an entirely different carburetor or injection system is required for supplying an engine with a gaseous fuel, propane carbs and injectors are designed for the leaner stoichiometric ratio. Propane engines operate more efficiently than gasoline engines, the difference being most pronounced at lower rpm.

Because propane enters the intake manifold as a gas, there is no cold-start or warm-up enrichment required; propane-fueled engines do not need a choke. Similarly, sudden acceleration requires no enrichment, unlike gasoline engines in which sudden drops in manifold vacuum on acceleration markedly decrease the vaporization of gasoline, hurting its ability to provide an appropriate air-fuel mixture. No accelerator pump-type device is required on propane carburetors.

Propane typically requires additional spark advance at low rpm due to its lower flame speed and decreased flammability. About a five degree increase is typical. At higher rpm, propane engines require less spark advance than gasoline-fueled engines—a total of 30 degrees, or less. Conventional distributors require different auto-advance springs for propane, and computers must be re-programmed with a different timing map. Sometimes Gann plates are installed on points-type distributors to limit advance on propane vehicles, and sometimes the slots of an advance plate and/or vacuum arm advance are welded, to limit advance. Dual-fuel gasoline-propane vehicles require a means for two entirely separate spark advance curves. This is easy on a computer but usually requires an add-on electronic "black box" on points-distributor engines.

Propane vapor increases the secondary voltage required to bridge a spark plug gap under all circumstances compared to gasoline vapor, and the requirements are particularly severe on sudden acceleration at low speeds. In general,

540-cubic-inch Donovan Chevy big-block motor uses two Turbonetics "Super" turbos and requires four propane liquid-to-gas converters and two mixers to supply fuel for 1,000 horsepower on propane and nitrous.

Cadillac dash supplies fuel and nitrous pressure data plus turbo boost. Aircraft-type circuit breakers protect circuitry.

Seville trunk is entirely filled by propane and nitrous tanks.

the condition of the ignition system is vital on propane-fueled vehicles. Some experts recommend reducing the plug gap from the 0.045–0.080 inch typical on gasoline-fueled vehicles to 0.035, or even less on systems originally equipped with plug gaps less than .040. Due to the lack of intake cooling, heavy-duty propane engines should run one heat range cooler plugs than gasoline-fueled versions of the engine. Normal plug carbon condition in a propane-fueled vehicle is indicated by a small amount of white carbon on the ceramic insulator of the plug. Burned electrodes indicate wrong mixture, wrong timing or advance curve, and too-hot plugs. Colder plugs require more voltage to fire, and propane-fueled vehicles running too-cold plugs may misfire on hard acceleration. Increased secondary voltages are strongly recommended from propane engine conversions. Experts suggest 8 milliliter plug wires should have a max resistance of 12,000 ohms per foot, with 7 milliliter plug wires running a max per foot of 8,000 ohms. Adjacent-firing cylinders should never have the plug wires routed next to each other to avoid cross-firing. Reversed coil polarity requires plugs to fire from the colder side electrode to the hotter center electrode, which requires more voltage. It is more critical on propane engines to avoid incorrect polarity.

Propane conversions of engines with computer engine management and catalytic convertors require special consideration. Computer-controlled engines usually have exhaust gas oxygen (EGO) sensors for air-fuel mixture calibration based on residual oxygen in the exhaust gases. Do not disconnect any of the system components. A disconnected air pump might affect the EGO sensor value. The computer also might turn on the check engine light with some components disconnected. Changing the thermostat to a cooler one for propane use might cause the computer to operate continuously in warm-up mode while running on gasoline. IAC idle speed stabilization devices operate on fuel-injected cars by allowing additional air to bypass the throttle plate; such devices must be reworked to draw additional air through the propane carb in order to avoid serious leaning if a propane carb is added for dual-fuel operation.

Engine Design and Tuning for Natural Gas

Propane and natural gas are very similar fuels in terms of performance characteristics and engine timing, etc. For motor vehicle use, the main practical difference is

The 540-cubic-inch motor uses custom coolant rail in foreground to route hot water to all four converters.

Propane Donovan engine on its way to Kim Barr Racing Dyno in Dallas.

that natural gas must be compressed or liquified, so vehicle storage tanks must operate under much higher pressures than with propane. Natural gas has lower energy content (but a higher octane rating) than propane, so natural gas vehicles usually can only travel 150–200 miles before refueling.

Both fuels benefit from high-powered capacitive discharge multispark systems like those from MSD. Like propane, natural gas requires modified ignition timing, which is complicated by the fact that the reduced range of most natural gas vehicles means that many of them have dual-fuel setups that require ignition systems that can provide optimal timing for both gasoline and natural gas. Companies like MSD (Autotronic Controls) make dual curve ignition systems with entirely different timing curves for different fuels. Natural gas requires more timing advance than gasoline at idle and lower rpm ranges, but the same total advance at high speeds. Gasoline fuel might knock at lower speeds with timing purely optimized for natural gas.

The optimal compression ratio for natural gas is 16.5:1, and the ideal stoichiometric mixture is 16.5:1. The CO level at stoichiometry will be about 1 percent, with slightly lower levels at cruise compared to idle for best drivability, and with full throttle CO levels at 3 percent for best power.

Dual-fuel natural gas engines typically make 15 percent less power on natural gas. Engine emissions and wear are typically drastically reduced, with CO and HC emissions reduced as much as 90 percent. NO_x emissions have been reduced 70 percent. Valve wear is not significantly worse where hardened seats are installed. Oil change intervals have typically been doubled compared to gasoline.

Laws on Conversion

Federal law allows a tax deduction of $2,000 per car or $50,000 per heavy-duty truck for converting to clean-burning fuels. Companies providing the conversion package should be familiar with EPA memo 1A, which specifies the EPA's position on the legality of conversions. Catalysts and other emissions control systems must remain in place on a vehicle. EPA has been working on emissions standards for converted vehicles, but any conversions done before the standards are in place will not have to meet standards. However, the EPA recommends following California Air Resources Board test procedures for conversions, in lieu of federal testing standards. EPA can be reached at (800) 423-1DOE, and CARB at (916) 322-2990.

Twin-turbo propane engine optimistically had only two converters prior to engine dyno session. Propane is opposite of gasoline in that its rich mixtures really heat things up.

11

Diesel

In the late 1880s, German scientist Dr. Rudolph Diesel conceived the idea of a sparkless engine in which the heat of compression rather than an electrical spark caused fuel to ignite. Diesel's prototype engine was designed to inject coal dust into the combustion chamber with compressed air. Independently, English engineer Herbert Akroyd-Stuart had already developed an "airless" liquid fuel-injection sparkless engine, but the Akroyd-Stuart engine relied on a heated metal "vaporizer" to initiate combustion as fuel splashed against it, rather than ignition by the heat of compression, so Diesel is given credit for inventing the modern compression-ignition engine.

Both the diesel and spark-ignition engines developed in the early 20th century, but diesel progress was slow until after WW1. Heavy, slow-turning diesel engines became more common for industrial, railroad, and marine applications. Diesel adaptation to automotive and aviation applications was far more limited. German gasoline shortages following the war led to more highly evolved diesel engines from Daimler-Benz and M.A.N. for commercial road vehicles, and interest spread to other European countries. By the mid-1930s, large numbers of diesel-powered vehicles saw service in road transport applications, for both goods and passenger transport, where diesel's advantages in fuel economy and lugging ability were clearly apparent. Enthusiasts tried diesel engines successfully in rally and racing applications, but diesels saw very limited acceptance in automobiles. An important application of diesel technology was in the German military.

With evolution of the early, slow-turning, heavy diesel engines into higher speed, higher output powerplants, higher performance diesel fuels became increasingly important. The first challenge was to tighten up diesel fuel quality, eliminating fuels with high viscosities or high levels of hard combustion residues. This allowed higher engine speeds, with higher power levels and improved reliability and fuel economy. The next challenge was to improve fuel ignition characteristics. As speeds increased by a factor of as much as ten, fuels that had been performing satisfactorily in slow-speed engines began having problems with combustion noise and startability.

The Germans not only built large numbers of diesel-powered tanks and trucks, they even built some bomber aircraft powered by

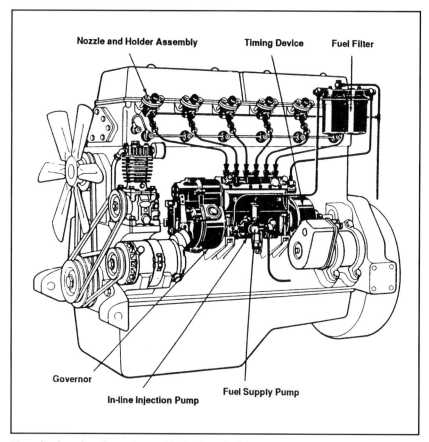

Six-cylinder diesel engine with in-line fuel-injection pump. Power level on diesels is based on the amount of fuel injected, since diesels normally have no throttle. Air is only a limitation when the engine is exhausting black smoke. *Robert Bosch Corp.*

Junkers Jumo diesel engines. While Allied armies standardized on gasoline as a military motor fuel, Wehrmacht armored divisions enjoyed the advantages of less-flammable diesel in armored combat vehicles. When struck, German tanks might continue operating, while Allied vehicles with similar wounds exploded in a ball of fire. Heavy diesel vehicles had other advantages. Diesel engines were more efficient, with no losses from throttling inlet air as is the case with gasoline engines, with the advantages of extremely high compression, and with the greater energy content per gallon of diesel fuel. Diesel engines had the additional advantage for heavy vehicles of extremely good lugging characteristics—the ability of diesel engines to make increased torque as they slow down.

In the United States, cheap domestic petroleum suppressed the adoption of the more economical diesel engine for heavy commercial road transport, but production of heavy diesels from Cummins, General Motors, and Caterpillar began in the 1920s for railroad, earth-moving, marine, and other off-highway uses. Since WWII, diesels have replaced gasoline engines in all heavy bus and long-haul motor freight applications. By the mid-1990s, government pressure to improve diesel emissions was resulting in significant incentives to convert diesel fleets to "clean-burning" natural gas, and the future of diesel engines is more uncertain than at any time in the recent past.

Diesel cars in the U.S. still represent a tiny fraction of motor vehicles. Diesel cars and light trucks have good fuel economy, but they are under-powered and noisy, and fuel supply remains problematic. Large increases in the fuel-efficiency of gasoline automobiles has somewhat decreased diesel's fuel economy advantages. The increasing popularity of light trucks in the 1980s and 1990s led to a growing aftermarket for high-performance modifications for them—primarily upgraded high-flow exhaust systems and turbocharger kits. There have been commercially available kits to add propane injection to heavy diesels as a power booster.

In the meantime, U.S. car companies with diesel automobiles experimented with building one-off race vehicles for publicity reasons, and have campaigned land speed-record vehicles at Bonneville and elsewhere. There has been much development of light, high-speed, indirect-injection diesel engines more suitable for automobile use. The latest trend has been converting diesel engines to run on alternative fuels with improved emissions characteristics, which has usually involved addition of a spark-ignition to a diesel cylinder head. There has been some experimentation with diesel engines for aviation.

Diesel Fuel Characteristics

Like gasolines, diesel fuels are mixtures of paraffinic, olefinic, naphthenic, and aromatic hydrocarbons, only with larger molecules than those in gasoline. Paraffins have the best startability but can cause trouble in cold conditions due to wax formation. Aromatics have opposite properties, with harder starting but better cold behavior—spontaneous detonation is the very characteristic needed to ignite diesel!

Cetane Number

The cetane number is a measurement of the tendency of diesel fuel to ignite when sprayed into a diesel engine—the higher the number, the easier the fuel ignites. Like gasoline's octane rating, cetane rating compares the auto-ignition characteristics of the test fuel to a reference blend of two fuels with a known cetane number: Normal cetane (n-hexadecane), with a defined cetane number of 100 and a

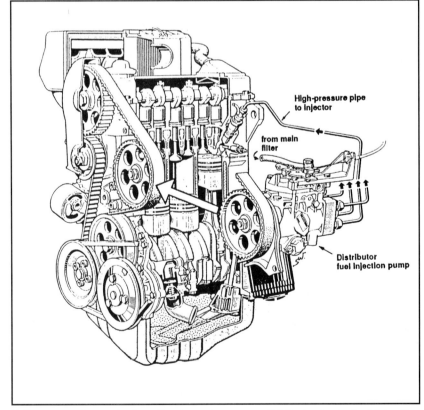

Volkswagen four-cylinder diesel engine uses distributor-type fuel-injection pump. *Volkswagen A.G.*

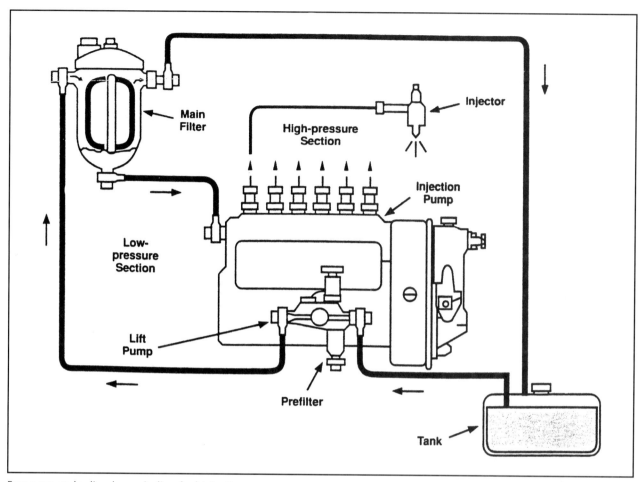

European-style diesels use in-line fuel-injection pumps.

high capacity to auto-ignite, and heptamethyl nonane, a highly branched paraffin with an assigned cetane value of 15 and poor auto-ignition characteristics. Cetane number is the percentage n-cetane plus 0.15 times the percentage heptamethyl nonane.

Most U.S. commercial diesel fuels have a cetane rating in the 40–51 range, which is somewhat lower than cetane standards in Europe and Japan, partly because a higher demand for gasoline in the U.S. forces refiners to use harsher reforming techniques to balance the naturally occurring components of crude oil with market demands, resulting in lower cetane blending components available for diesel fuel. Running an engine on a lower than design cetane fuel causes harder starting and raises peak

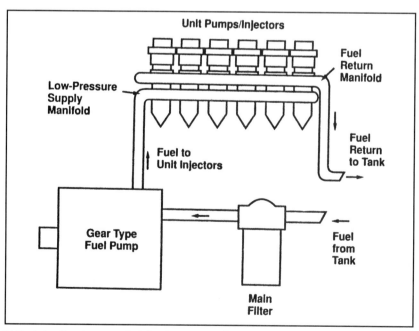

U.S.-style diesels use unit injection pumps/injectors.

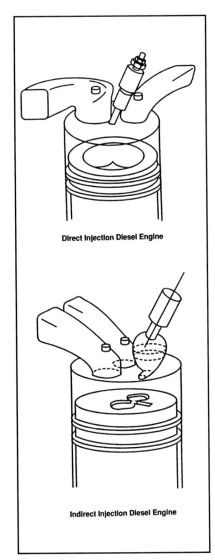

Two types of diesel combustion chambers.

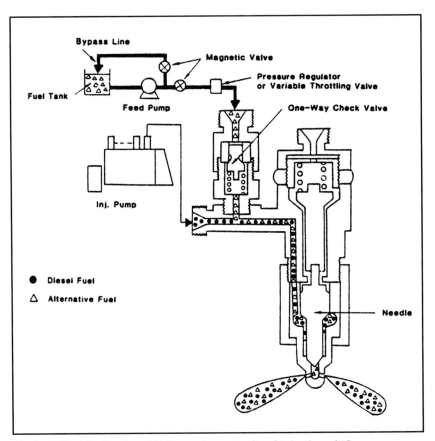

Alcohol diesel fuel system (SAE) supplements diesel injection. *SAE*

cylinder pressures, combustion noise, and emissions, since more fuel will be injected before ignition and less will have burned before the exhaust valve opens. Higher cetane fuels ignite sooner, generally reducing combustion noise and emissions, although a really high cetane fuel could ignite before sufficient air-fuel mixing took place, raising emissions and lowering power if early burning causes a big rise in cylinder pressure before the piston passes TDC. As with spark-ignition engines' ignition timing, and its relationship to fuel octane rating, diesel injection timing is critical, and must be adjusted for out-of-spec cetane fuels.

Although harder starting and noisier, lower cetane diesel fuels actually have a higher heating value, so there is a trend to lower fuel consumption with lower cetane.

Besides ignition quality, the most important fuel characteristics affecting diesel combustion are density, volatility, and viscosity. Density influences the heating value of the fuel, which determines its ability to make power. However, for a diesel engine to make best power, the fuel must lend itself to

Exhaust Emissions	Methanol Diesels		Typical Diesels
	MAN	DDC	
HC, g/km	0.53	75.0	1.7-2.5
CO, g/km	0.48	55.0	10-18
NO_x, g/km	8.80	4.9	14-19
Particulates, g/km	0.06	0.96	1.7-3.9
Aldehydes, g/km	0.10	1.20	not detected
Methanol, g/km	0.35	62.0	not detected

Emissions from methanol and conventional diesels.

precise metering, filtering, and atomization, which are influenced by the above characteristics plus cold flow, deposit-forming tendency, cleanliness, and corrosiveness.

Since diesel injection equipment meters fuel based on volume, changes in density impact engine power output due to the varying mass of fuel injected. Higher density fuels will produce more power and tend to produce more smoke if the engine is tuned close to the smoke limit. Power output has been shown to increase from roughly one-half percent to over 1.5 percent with higher density fuels.

High-Performance Diesel Systems

Unlike a spark-ignition engine, a diesel engine normally has no throttle, breathing in a full charge of air no matter what the power output. Fuel is injected at extremely high pressure directly into the combustion chamber near the top of the compression stroke and spontaneously ignites in the superheated compressed air after a very short delay, while fuel is still being injected. Diesel engines are designed with compression ratios up to 22:1, both to provide the heating required for combustion, and because they can utilize these efficient high compression ratios without fear of explosive detonation. Power output of a diesel engine is thus entirely dependent on the fuel injection system, and it is common for diesel engines to be running with an air surplus in order to avoid smoking. Accordingly, turbocharging a diesel engine requires that the injection pump be recalibrated to deliver more fuel. Competition diesel pulling tractors are normally tuned such that black smoke pours out of the exhaust under full power, making sure there is fuel present for every last molecule of air drawn or forced into the engine. Manifold pressures as high as 20 atmospheres have been successfully utilized in two- or three-stage turbo systems on such machines.

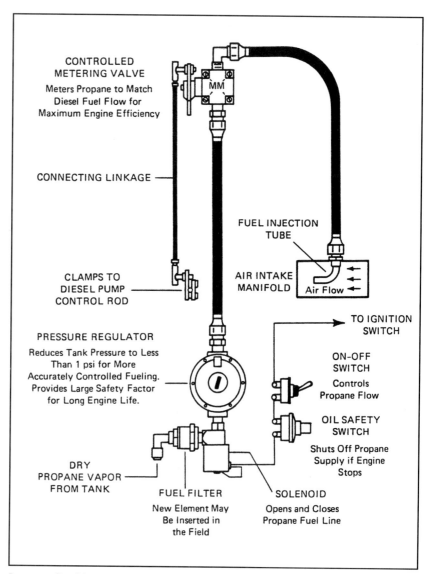

Propane diesel boost system makes use of air not used for diesel combustion to burn auxiliary propane fuel for extra power with no black smoke. *Reston Publishing Co.*

Turbocharging or supercharging diesel engines is the accepted method of improving performance, although rich mixtures can have a beneficial effect where the engine is running with a surplus of air, and propane injection is beneficial where the injection capacity is limited. Turbocharging a diesel engine does more than improve the VE of the engine, improving combustion efficiency by increasing combustion chamber turbulence, which provides more complete combustion.

Turbocharged diesels typically run a 50 percent surplus of air at sea level, and with electronic wastegates can maintain full VE to very high altitudes, depending on the capacity of the turbocharger. Turbo diesels typically have an EGT 150–200 degrees lower than naturally aspirated ones.

Too-early injection causes combustion pressure to peak before TDC, causing power loss and possible engine damage; injection too late causes noise, power loss, and possible exhaust valve damage as the still-burning fuel exits the combustion chamber. EGT goes up while power goes down. Diesel in-

Engine	Cat 1A	Cat 1H2	Cat 1D	Cat 1G2	PETTER AV1
Bore mm (in.)	146.0 (5¾)	130.0 (5⅛)	146.0 (5¾)	130.0 (5⅛)	86.0 (3⅜)
No. of Cylinders	1	1	1	1	1
Speed (rpm)	1000	1800	1200	1800	1500
Supercharge kPa (ins. Hg.)	nil	33.86 (10)	50.79 (15)	77.89 (23)	nil
Test Duration (h)	480	480	480	480	120
B.M.E.P. kPa (lbs/in.²)	524.0 (76)	758.4 (110)	903.2 (131)	965.2 (140)	537.8 (78)
Fuel Sulphur Content, % Wt.	0.35 min. or 0.95-1.05	0.35 min.	0.95-1.05	0.35 min.	0.35 min. or 0.95-1.05
Sump Temp. (°C)	65	82	80	96	55
Coolant Temp. Outlet (°C)	80	71	93	88	85 (Kerosene)
Oil Change Period (h)	120	120	120	120	None
B.H.P.	20	33.5	42.0	40.0	5.0
Top Ring Groove Temp. (°C)	208	220	220	243	220
H.P. per L (gal.) Sump Charge	60.6 (16)	101.4 (26.8)	127.2 (33.6)	130.2 (34.4)	30.3 (8.0)
MIL-L Specification	2104A or Supp 1	46152	2104C	2104C	*DEF STAN 91-43

*British Specification—No test oil added during test.

Categories for diesel engine lubrication. Diesel engines require different lubrication characteristics than gasoline engines, although oils may meet both specs. *Lubrizol*

jection systems must advance injection timing with increased engine speed, exactly as a spark-ignition engine advances spark timing, and naturally, the injection pump and injector must be sized so as to provide enough fuel soon enough for it to burn in the time available. Diesel injection systems with very short injection time clearly have the ability to be calibrated for increased power if the air is available. If the injection time is long, the only option may be auxiliary fueling like propane injection.

Turbocharging expert Hugh MacInnes has specified an arbitrary EGT limit of 1,300 degrees for increasing diesel power by turbocharging. If EGT rises without a power increase, injection timing is probably retarded, but injection timing should not be advanced beyond naturally aspirated limits, to avoid exceeding safe combustion pressures.

Antiknock Additives and Water Injection

12

There are several types of antiknock additives available. This chapter discusses the most common ones.

TEL

The most effective octane-booster for gasoline is tetraethyl lead, an organo-metallic compound thought to work by forming lead oxide during combustion, which then collects free radicals and prevents them from initiating knock-producing chemical reactions. TEL is so deadly in pure form that it must be handled by licensed handlers wearing rubber gloves, full masks, and rubber suits. It was never available to the public.

The first gram of lead added to a gallon of gasoline raises the (R+M)/2 (pump octane) about six octane numbers. There are diminishing returns with additional lead, but up to 4–5 grams of lead per gallon improve octane as much as 10 or 11 points. Improving the octane of low-octane gasoline components with lead additive is cheaper than producing high-octane components, which consumes more crude oil.

Both TEL and TML (tetramethyl lead) are very effective octane improvers for gasoline but are not effective for other fuels. Alcohols and the gaseous fuels have octane ratings sufficient for the most severe conditions; nitromethane, which has a high tendency to knock, is not helped by TEL or TML.

The conversion to unleaded street fuels was precipitated by the discovery that TEL coats and ruins catalytic converters. TEL was entirely eliminated from street gasoline in the U.S. in 1995, although it is still legal for aircraft and off-highway use, and leaded racing gasolines have been produced using up to 5 grams of TEL per gallon, resulting in octane ratings 20 points higher than street gas. In high concentrations, lead can foul plugs, so lead-scavenging additives containing bromine and chlorine have been added to remove lead deposits from the combustion chamber. Organic residues from burned oil or residual fuel can bind to residual lead, defeating the scavenging effect of the additives.

Although TEL is still legal for racing gasolines for off-highway-only use, tests indicate that high-octane unleaded fuels are as good or better in performance terms. Racers buying custom-mixed leaded gasoline should treat leaded gasoline like the hazardous material that it is—TEL exposure gives children learning disabilities and makes adults stupid. Lead can be absorbed through the skin. Stay away from it.

MMT

Methylcyclopentadienyl manganese tricarbonyl (MMT), another organo-metallic compound, has been used mainly to enhance the effects of TEL in improving octane. It has also been used in quantities up to 1 gram per gallon as an antiknock additive for racing gasolines, both by itself and in combination with TEL and TEL-extender acetic acid. Oddly enough, acetic acid lowers octane in the presence of MMT without TEL. Like TEL, MMT can cause plug and valve deposits. Although toxic, MMT is available as an aftermarket octane booster in diluted form.

Aniline

Large quantities of aniline are necessary to have much of an effect on boosting gasoline octane rating; it has only two or three percent the effectiveness of TEL. It is expensive, so it has been used mainly by racers trying to upgrade pump gas for competition. Aniline is highly toxic.

Ethanol

As an octane booster, ethanol has a MON of 90 to 110, depending on the compounds in the base gasoline, and has been available commercially as an aftermarket octane booster for competition use. According to SAE paper 852129, ethyl alcohol can increase the specific energy enough to make it illegal for racing under certain rules. Its presence in gasoline can be detected by the increased dielectric constant of the blend.

Methanol and MTBE

As an additive, methanol is similar to ethanol, although more corrosive and dangerous. MTBE, an ether, contains oxygen like methanol and ethanol, but in a form which makes it a safer and more convenient fuel blending agent.

Water Injection

In 1913, adding water to an engine was proposed by Professor Hopkinson of Cambridge as an alternative to a separate cooling system, and indeed, studies found it was possible to eliminate the cooling system, although the mass of water required to do this was several times that of the fuel, and the water adversely impacted combustion quality.

Engineers have always understood that increased compression resulted in increased thermal efficiency, and given finite octane quality, water injection has been used to

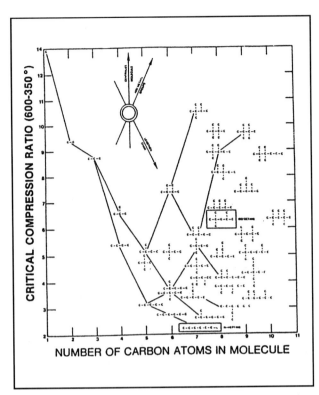

Critical knock-limited compression ratios for various hydrocarbon fuels. *SAE*

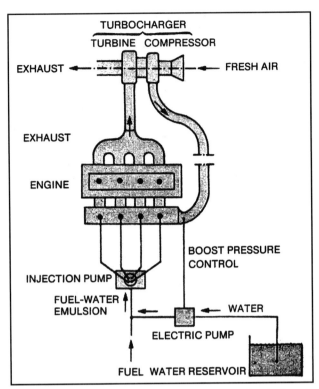

Water-injection system in which water is mixed directly with fuel as an emulsion prior to injection.

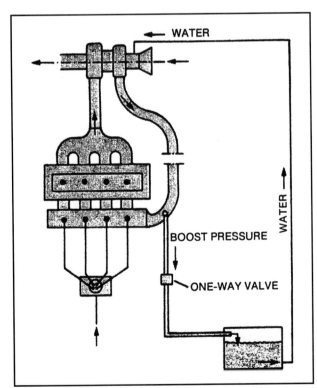

Water-injection system in which water is sprayed into turbocharger inlet air and distributed to cylinders via air pressure.

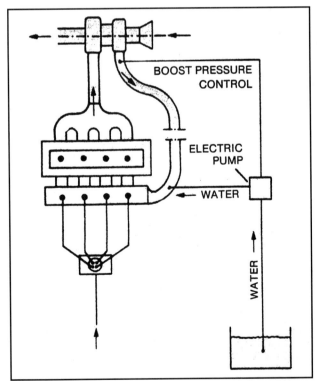

Water-injection system in which water is injected directly into the intake manifold; it is entirely independent of the fuel-injection system.

	Blend #	Furan 73	Tetra-hydrofuran 30	Propylene oxide 46	Dioxane 72	Nitro-methane 220	Nitro-propane 221	Isoprene 74	Styrene 37	Methylene chloride 122
Volume %:										
Alkylate		55	50	50	55	55	55	55	50	55
Toluene		25	25	25	25	25	25	25	10	25
Compound		20	25	25	20	10	10	20	15	20
MTBE		–	–	–	–	–	–	–	25	–
Acetone		–	–	–	–	10	10	–	–	–
	mg Mn/L				Octane Results					
RON	0	108.6	95.2	106.3	93.3	99.0	96.6	104.4	111.8	106.1
MON	0	94.5	85.6	94.0	87.6	75.4	71.9	92.0	96.0	101.4
(R+M)/2	0	101.6	90.4	100.2	90.5	87.2	84.3	98.2	103.9	103.8
RON	25	109.4			95.6			104.8		
MON	25	95.2			88.2			92.5		
(R+M)/2	25	102.3			91.9			98.7		

Antiknock properties of various compounds in alkylate, toluene, and MTBE, alone or with Mn. *SAE*

suppress spark knock by slowing down the rate of combustion.

WWII fighter aircraft sometimes used water or water-alcohol injection to increase the manifold pressure that could be tolerated under all-out combat conditions.

In modern times, hot-rodders have used water injection as an antiknock strategy, particularly on high-compression engines from the 1960s when high-octane gasolines were no longer available. It has also been used on turbocharged street vehicles to increase the boost that could be run without detonation.

Water Injection Studies

GM Research studied the impact of water introduction on the combustion process, both in the fuel as an emulsion and directly, via injection into the intake air, and discovered the following:

Water injection was greatly preferable to the emulsion system, mainly due to the pro-knock characteristics of the emulsifier.

Adding water at a rate of up

	Blend #	Methyl Acetate 172	Ethyl Acetate 48	Ethyl Acetate 44	n-Propyl Acetate 45	p-Amyl Acetate 47	Bomyl Acetate 179	Ethyl Butyrate 177	Butyl Benzoate 178	Diethyl Carbonate 180
Volume %:										
Alkylate		50	50	50	50	50	55	50	60	55
Toluene		15	15	25	25	25	30	30	30	25
Ester		35	35	25	25	25	15	20	20	20
	mg Mn/L				Octane Results					
RON	0		112.1	110.1	108.8	102.7	105.4	107.9	105.2	108.7
MON	0		103.3	99.9	99.8	95.7	96.8	97.5	96.0	98.6
(R+M)/2	0		107.7	105.0	104.3	99.2	101.1	102.7	100.6	103.7
RON	25	112.7	113.6	111.4	110.4					
MON	25	104.7	105.1	101.7	101.2					
(R+M)/2	25	108.7	109.4	106.6	105.8					
RON	50		114.6							
MON	50		106.2							
(R+M)/2	50		110.4							
Density		0.9412	0.7961	0.7932	0.8936					
RVP psi		7.1	2.5	2.3	1.2					
Net Heat of Combustion	MJ/Kg	19.7	36.5	38.2	26.6					
	Btu/lb	8,479	15,669	16,404	11,421					
wt % Oxygen		15.1	12.7	9.1	7.8					
Stoichiometric Air/Fuel Ratio		11.9:1	12.3:1	13.1:1						

Antiknock response of various esters with and without Mn. *SAE*

	Blend #	Alcohols				Ketones			Amines		Diethyl-amine
		Methanol 102	Methanol 103	Ethanol 126	Isopropanol 100	Acetone 118	MEK 119	MIBK 36	Aniline 49	N-Methyl Aniline 81	
Volume %:											
Alkylate		65	55	65	65	50	55	40	65	65	55
Toluene		25	25	25	25	25	20	30	25	30	20
Compound		5	5	5	–	25	25	30	10	5	5
Isopropanol		5	5	5	10	–	–	–	–	–	–
MTBE		–	10	–	–	–	–	–	–	–	20
Octane Results	mg Mn/L										
RON	0	107.1	108.7	107.6	106.8	109.0	108.6	109.6	116.6	110.4	104.0
MON	0	96.9	97.0	95.5	97.3	100.0	99.1	99.9	105.6	96.2	92.3
(R+M)/2	0	102	102.9	101.6	102.1	104.5	103.9	104.8	111.1	103.3	98.2
RON	25	110.1	108.9	109.4	108.2	110.2	109.9	110.6	116.4	110.8	
MON	25	97.5	97.2	97.2	98.2	101.8	100.0	100.2	105.3	96.6	
(R+M)/2	25	103.8	103.1	103.3	103.2	105.9	105.0	105.4	110.9	103.7	
Density		0.7522	0.7561	0.7519	0.7516	0.7659	0.7606	0.7863	0.7755		
RVP psi		5.4	6.0	3.6	3.4	3.4	3.3	3.3	1.9		
Net Heat of Combustion MJ/Kg		41.5	40.6	41.8	42.0	32.5	40.4	40.4	42.5		
Btu/lb		17,830	17,436	17,978	18,057	16,958	17,341	17,343	18,242		
Wt % Oxygen		3.8	4.7	3.1	2.7	6.9	5.8	4.8	0		
Stoichiometric Air/Fuel Ratio		14.1:1	12.6:1	14.2:1	14.3:1	12.8:1					

Antiknock properties of alcohols, ketones, and amines. *SAE*

to 40 percent by weight increased the research octane linearly up to roughly 10 numbers, from 90 to 100. Motor octane increased in a linear fashion from 82 to 87 at 20 percent water. (Due to the heat of vaporization of the water, it was not possible to obtain a motor octane number with 40-weight water addition.)

ALKYLATE/MTBE/TOLUENE BLEND	MMT	FERROCENE	IRON CARBONYL
RESEARCH OCTANE NUMBER			
Clear	108.2	108.2	108.2
mg metal/L	Manganese	Iron	Iron
12.5	109.2 (10mg)	109.1	–
25	109.7	109.4	–
50	110.8	110.1	110.0
75	–	110.6	–
100	111.3	111	110.8
150	–	–	111.7
25 mg MMT	109.7		
25 mg 50/50 MMT/FERROCENE		109.8	
25 mg 50/50 MMT/CARBONYL			109.4
50 mg 50/50 MMT/FERROCENE		110.2	
MOTOR OCTANE NUMBER			
Clear	97.3	97.3	97.3
mg metal/L	Manganese	Iron	Iron
12.5	98.0 (10mg)	98.4	–
25	98.7	98.6	–
50	99.1	99.1	98.3
75	–	99.2	–
100	99.8	99.3	99.2
150	–	–	99.7
25 mg MMT	98.7		
25 mg 50/50 MMT/FERROCENE		08.8	
25 mg 50/50 MMT/CARBONYL			98.4
50 mg 50/50 MMT/FERROCENE		98.5	

Metallic antiknock compounds compared. *SAE*

Alternatively, without increasing fuel octane, 40 percent water addition allowed the knock-limited engine compression ratios to be increased one full ratio (in the testing, from 8:1 to 9:1). This essentially will increase engine efficiency by three percent. Other than this, the effects of water injection essentially balanced each other out. The reduced compression workload due to water's heat of vaporization reduced gas pressure by lowering temperature; conversely, the reduced peak combustion pressure decreased the work done during the power stroke.

Minimum spark advance for best torque (MBT) increases with water injection; 40 percent water increased MBT requirements by 5 to 15 degrees.

The addition of water to the inlet charge had little or no effect on thermal efficiency, volumetric efficiency, lean operating limits, smoke, exhaust temperature, or engine cooling requirements.

Exhaust emissions were affected, with NO_x decreasing up to 40 percent with 40 percent water, while hydrocarbon emissions increased about 50 percent with direct manifold water

Blend #		334	1	166	167	6	27
Volume %							
Alkylate		45	50	60	70	55	55
MTBE		55	50	40	30	20	20
Xylene		–	–	–	–	10	–
Toluene		–	–	–	–	15	–
Cumene		–	–	–	–	–	25
			Octane Results				
	mg Mn/L						
RON	0		110.2	110.0	108.7	107.5	109.2
MON	0		99.2	99.3	98.6	97.2	97.8
(R+M)/2	0		104.7	104.7	103.7	102.3	103.5
RON	25	114.4	113.7	112.2	110.9	108.8	111.0
MON	25	100.0	101.8	99.7	99.7	98.0	99.1
(R+M)/2	25	107.2	107.8	106	105.3	103.4	105.1
RON	50	115.3	115.2	113.6	112.4		112.9
MON	50	100.3	101.7	100.7	100.3		99.3
(R+M)/2	50	107.8	108.5	107.2	106.4		106.1
Density		0.7231	0.7210	0.7168	0.7126	0.7514	0.7505
RVP psi		6.2	5.9	5.2	4.5	3.5	3.3
Net Heat of	MJ/Kg	39.3	39.8	40.7	41.6	41.6	41.7
Combustion	Btu/lb	16,894	17,092	17,486	17,881	17877	17937
wt% Oxygen			10.0	9.1	7.3	5.5	3.6	3.6
Stoichiometric Air/Fuel Ratio		13.3:1	13.4:1	13.8:1	14.1:1	14.0:1	14.1:1

Antiknock properties of alkylate/MTBE blends and response of xylene and cumene. *SAE*

Blend #		20*	97	22	66	67	26	12
Volume %:								
Alkylate		70	55	50	45	35	25	40
Toluene		30	25	25	25	25	25	30
MTBE		–	20	20	30	40	50	30
Butane		–	–	5	–	–	–	–
				Octane Results				
	mg Mn/L							
RON	0	105	108.5	107.5	108.6	109.2	111.4	108.9
MON	0	97.5	97.8	97.0	97.9	97.9	97.8	97.8
(R+M)/2	0	101.3	103.2	102.3	103.3	103.4	104.6	103.4
RON	25	106.2	109.5	109.2	110.8	111.6	112.7	111.0
MON	25	98.6	98.6	98.2	98.4	98.9	98.9	98.0
(R+M)/2	25	102.4	104.1	103.7	104.6	105.3	105.8	104.5
RON	50			110.5		111.7	113.3		111.9
MON	50			99.1		98.9	99.1		98.6
(R+M)/2	50			104.8		105.3	106.2		105.3
RON	100			111.3		112.6	114.3		
MON	100			99.8		99.5	99.5		
(R+M)/2	100			105.6		106.1	106.9		
Density		0.7516	0.7514	0.7456	0.7556	0.7598	0.7639	0.7642
RVP psi		2.1	3.5	6.3	4.1	4.8	5.5	4.1
Net Heat of	MJ/Kg	43.2	41.6	41.6	40.7	39.7	38.8	40.5
Combustion	Btu/lb	18,572	17,865	17894	17,470	17,076	16,681	17,388
wt % Oxygen		0	3.6	3.6	5.5	7.3	9.1	5.5
Stochiometric Air/Fuel Ratio		14.6:1	14:1	14:1	13.7:1	13.4:1	13:1	13.6:1

*Primary Reference Blend

Antiknock properties of MTBE blends. *SAE*

injection, although the water was not decreasing the efficiency of the catalytic converter. CO emissions essentially stayed the same.

Water injection tends to contaminate the lubrication system. This can reportedly be eliminated by direct injection of water into the cylinder near the end of compression. Water injection may corrode internal engine surfaces, and in cold climates, injection water freeze protection can be achieved by the addition of methanol.

Spearco and others offer water injection kits for various types of turbocharged and naturally aspirated engines. Typically, water is injected into the throat of a throttle body, carb, or turbocharger.

	RESEARCH			MOTOR		
	Octane Number	Blending Values		Octane Number	Blending Values	
		*API Project 45	**Gasoline Blend		*API Project 45	**Gasoline Blend
Benzene		98	106	115	90	88
Toluene	120	124	114	104	112	93
p−Xylene	116	146	120	110	126	98
m−Xylene	118	145	120	115	124	99
o−Xylene	100	120	105	100	102	87
Ethylbenzene	107	124	114	98	107	91
Isopropylbenzene (Cumene)	113	132		99	124	
n−Proplybenzene	111	127		99	129	
C9 Aromatics			117			98
C10 Aromatics			110			92
Reference	(4)	(4)	(6)	(4)	(4)	(6)

* 20% aromatic 80% 60:40 mixture isooctane/n−heptane
** 20% aromatic 80% refinery grade gasoline

Aromatic octane comparison. *SAE*

Aftermarket Octane Booster Products

VP Racing Fuels
C5 Octane Booster

VP claims C5 raises octane up to eight numbers. De-carbonizer additives are claimed to lower engine octane requirements and fight pre-ignition. A lead lubricant is included to protect the engine and extend its life. VP claims this is the only additive that "puts back what you've been missing at the pump." Each bottle treats 20 gallons of gasoline. C5 is not for use in catalyst-equipped vehicles.

104+ Octane Booster

Available at automotive aftermarket stores, 104+ is an over-the-counter additive designed to be street legal.

	Research Octane Number						Motor Octane Number					
	Clear	mg Mn/L					Clear	mg Mn/L				
		+10	+18	+25	+50	+100		+10	+18	+25	+50	+100
MTBE	114.6	115.5		116	116.3	117.4	98.5	99.7		99.9	100.7	101.8
Alkylate 1	97.2	98.6		99.5	99.8	101.9	97.9	98.6		99.9	100.7	102.2
Alkylate 2	94.6		96.0	96.6	98.4		94.0			95.5	97.1	
Alkylate	94.5						92.3					
Reformate	94.8						84.0					
Light Cat Cracked	96.6						83.9					
Isooctane	100.0	100.6		102.0	107.2		100.0	101.7		104.0	106.4	
Toluene	117.6	117.8		117.8								
Toluene Standardization Fuel (74% Toluene 26% Isooctane)	113.7	113.9		115.1			100.8	100.7		102.1		

Refinery stream hydrocarbon octanes, neat or with MMT additive.

13

Nitrous Oxide Injection

Nitrous oxide was discovered in 1772 by English scientist Joseph Priestly. Nitrous oxide, N_2O, turned out to be a nontoxic, nonirritating clear gas with a slightly sweet taste and odor. Later on, another Englishman, Humphry Davy, utilized nitrous oxide in the Davy miner's lamp as a self-contained source of oxygen, which eliminated the threat of open-flame oil lanterns igniting flammable gases in mines. Along the way, Davy discovered that nitrous oxide was a mild anesthetic when inhaled with air in the right ratio (inhaling pure nitrous oxide will cause death by asphyxiation). In more modern times, nitrous oxide became commonly used in medicine for anesthetic purposes, especially in dentistry. When inhaled in small quantities, nitrous produces mild euphoria or hysteria and spells of giggling and laughter—hence the slang name "laughing gas."

Research with nitrous oxide as an oxidizer and power improver for high-altitude aircraft began following WWI and became a major research effort during WWII. It was clear by 1942 that a good fighting aircraft required high speed and the ability to operate at very high altitudes. British scientists began a massive secret effort to develop a nitrous system that could boost power and altitude capability without significantly impacting reliability, which eventually allowed reconnaissance aircraft to operate over Axis territory at altitudes so high they were essentially invulnerable. U.S. scientists did similar research on V-12 engines from Packard and Rolls Royce, boosting operational ceilings to well over 50,000 feet, although nitrous was never used significantly in combat aircraft. In the meantime, German scientists had developed their own nitrous injection system for the Daimler-Benz DB603 inverted V-12 as fitted to certain Messerschmitt 109 aircraft. Nitrous was injected directly into the inlet of a gear-driven supercharger, the combination of supercharging and nitrous achieving full rated horsepower at extremely high altitudes. At an injection rate of 12 pounds per minute, the nitrous oxide provided oxygen for 350 additional horsepower; injection could be maintained for up to 30 minutes.

The advent of turbojet aircraft made all the nitrous research obsolete for aircraft use, and nitrous was virtually forgotten in the post-war years. Although automotive motorsports publications made isolated references to nitrous injection in the late 1950s and early 1960s, those racers who knew about the potential of nitrous guarded their knowledge with secrecy. Early practitioners used crude nitrous injection systems disguised so as to keep them clandestine. It was not until the early 1970s that nitrous injection became common knowledge among hot-rodders and racers.

Nitrous appeared to be a hot-rodders dream-come-true: cheap, easy to install, easy to maintain, and effective in making significant horsepower gains. A rash of entrepreneurs jumped into the expanding market with nitrous kits, some of which were not reliable. A predictable rash

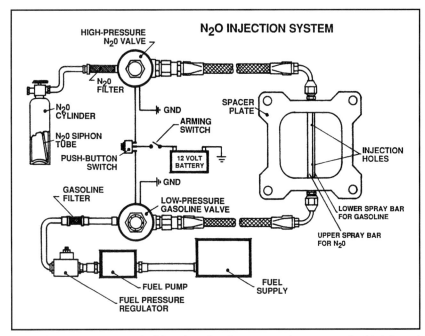

NO_2 injection system. Typical four-barrel carb-type spraybar system. Fuel is injected through the lower bar, NO_2 through the upper, always making sure oxidizing fuel doesn't reach the combustion chambers without enrichment fuel. *S-A Design Books*

99

of explosions and catastrophic engine failures followed. Natural selection and market forces eventually resulted in inferior kits being upgraded to high quality or the companies that made them going out of business.

Nitrous Functionality

Simply put, nitrous oxide is not a fuel but rather a source of additional oxygen for an engine, enabling more fuel to be burned, which in turn produces more power. Pure oxygen injection is considered dangerous because an oxygen leak could cause dangerous or even explosive combustion of virtually anything, from fuel to oil or grease to hot steel or cast-iron combustion chamber components. Perhaps most importantly, concentrated pure oxygen mixed with air produces dangerously hot combustion that can kill an engine in a matter of two or three seconds. Nitrous oxide injection is safer because the oxygen atom in nitrous is strongly bonded to two nitrogen atoms, so the oxygen is unavailable at ambient temperatures, and when dissociated above 565 degrees it is heavily diluted with nitrogen.

Nitrous oxide is a gas at standard temperature and pressure (70 degrees, 14.7 psi). Vapor pressure in a closed tank will hold nitrous in liquid form at 760 psi at standard temperature. Above 97.7 degrees F, nitrous will not stay a liquid, no matter what the pressure. It is a liquid at standard pressure at -128 degrees F. At temperatures below 97.7 degrees F, it is possible to extract either liquid or gaseous nitrous from the tank, depending where in the tank the pickup draws from. Liquid nitrous exiting from a tank will expand very rapidly due to the pressure drop from 760 to 14.7 psi, and will consequently undergo a substantial cooling, becoming super chilled. This effect can be used to cool inlet air and to keep combustion temperatures down, helping to offset nitrous' tendency to increase the likelihood of detonation.

Nitrous oxide is 36 percent oxygen by weight. The chemically correct nitrous-gasoline ratio is 9.6:1, and fuel enrichment for the nitrous injection is often handled separately from the normal fueling system, such as a spray bar under a four-barrel carb or additional fuel injectors, or at the very least,

Author's 12-injector six-cylinder Jaguar XKE uses Nitrous Oxides Systems' computer fuel injection with nitrous enrichment fuel supplied by lengthening injection pulse width upon nitrous activation. The EFI Technologies' computer is capable of pulsing the nitrous solenoid to bring in nitrous boost relatively gradually.

Nitrous oxide injection is common and legal on street cars like the late-model Corvette, providing one of the simplest and most effective avenues to a large power boost.

an independent calculation and addition to injection quantity for nitrous fueling. However, nitrous and fuel for nitrous are not actually entirely independent of air and normal fuel intake, because the nitrous (particularly gaseous nitrous) and enrichment fuel will displace air in the intake tract and so affect the total oxygen-fuel mixture ratio. In addition, the cooling effect significantly affects the density of the intake charge. These factors must be considered in computing the amount of nitrous fuel to provide.

The total fuel almost always includes an excess beyond the chemically ideal mixture, to help keep combustion temperatures reasonable. By the same token, lean mixtures can be disastrous in a nitrous motor. With no excess fuel to keep combustion temperatures down, hot spots can appear in the combustion chamber, which in turn can cause the onset of pre-ignition and detonation, further raising temperatures to the point where excess oxygen will attack the red-hot surfaces and burn them away exactly like an oxidizing acetylene torch.

Lean nitrous-fuel ratios must be avoided at all costs. With gasoline, the maximum horsepower in a normally aspirated engine occurs at air-fuel ratios 15–20 percent rich of stoichiometric. Nitrous motors burn a gas-fuel mixture with enhanced oxygen content but less nitrogen, so there is less inert nitrogen to control combustion temperatures, automatically causing combustion temperatures and flame speeds to increase.

NOS port injector is capable of injecting both fuel and gasoline at individual intake ports or at the inlet of a turbocharger.

Mathematics of Nitrous-Enhanced Combustion

A medium-sized V-8 engine running fast at heavy load might consume 400 cfm of air, or roughly 30.6 pounds of air per minute at standard temperature and pressure. Of this, 23.6 percent is oxygen, meaning the engine is actually consuming 7.2 pounds per minute of oxygen. Assume the vehicle has a nitrous system that delivers 5 pounds of liquid nitrous per minute, 36 percent of which is oxygen, then the nitrous system is delivering 1.8 pounds of oxygen and 3.2 pounds of nitrogen per minute. Adding the nitrous-oxygen to the air-oxygen, the nitrous-enhanced intake mixture contains roughly 9 pounds of oxygen, meaning slightly more than 25 percent of the air mixture is oxygen, and 26.5 pounds is nitrogen, or nearly 75 percent. Oxygen content has increased over seven percent while nitrogen has decreased almost 1.5 percent. If gaseous nitrous were injected, there are at least two consequences: The cooling effect of liquid nitrous boiling disappears, while the nitrous gas displaces air (and fuel), which would otherwise have produced power. Gaseous nitrous injection is likely to have hotter combustion temperatures. Even when injecting liquid nitrous, steps must be taken to control combustion temperatures, which could otherwise easily increase by a greater percentage than the increased oxygen content. One vital tactic is effectively managing intake charge temperatures.

Nitrous Fuel Enrichment-Fuel Ratios and Performance Considerations

The proper ratio between nitrous-oxide and enrichment fuel varies according to the engine and application and, of course, according to the fuel. In the case of gasoline, the specific gravity, BTU content, and oxygen content will affect the optimal nitrous-gas ratio. Using the pure reference fuel iso-octane to represent gasoline, the chemically correct ratio is 9.649. For nitrous-propane, the ratio is 10; for nitrous-methanol 4.13; for nitrous-ethanol 5.74; for butane 9.930; and for nitromethane, it's 1.090. Actual fuel mixtures should be increased well rich of stoichiometric to achieve maximum power and to provide combustion chamber cooling. In general, nitrous injection tends to magnify the weaknesses of any particular fuel in racing or other extreme conditions.

Gasoline and Nitrous

Street gasoline's relatively low-octane rating will be more of a problem with nitrous injection. A gasoline engine susceptible to knock will have worse knock problems with nitrous. Combustion temperatures must be controlled. High-octane racing fuel for all fueling or a dual-fuel system with high-octane fuel for nitrous enrichment can be effective in handling knock. For vehicles that must run on street gasoline, richening the mixture or water injection are effective in controlling combustion temperatures. Any mechanical means to improve combustion chamber quench by improved head cooling or oil splash cooling of pistons will help.

Alcohol and Nitrous

Both methanol and ethanol make maximum power very rich of stoichiometric mixtures, and methanol in particular has tremendous cooling properties when used in a fuel-rich mixture, which is valuable in fighting detonation and increasing engine volumetric efficiency. Alcohol fuels are subject to pre-ignition problems, and nitrous injection exacerbates this problem. Combustion chamber hot spots on the verge of pre-igniting the mixture are heat-

1972 Pantara "Megabuck," designed for Bonneville competition, used a twin-turbo SOHC 427-cubic-inch Ford engine to make 1,500 horsepower.

Megabuck's engine compartment shows port injection lines (between the twin compressor outlets, just behind the plenum atop the engine), which were originally designed for methanol injection.

ed further by nitrous combustion, causing pre-ignition.

Nitrous expert David Vizard has suggested that nitrous systems for small-block V-8s, properly recalibrated, will deliver 50–80 more horsepower, and a big-block V-8, 80 to 100 more horsepower. In some ways, alcohol and nitrous are made for each other: Nitrous heats up combustion, alcohol cools it down. Nitrous lowers the detonation threshold, alcohol resists detonation very effectively. Alcohol has a higher specific energy, and its rich flammability limits and ability to run on rich mixtures tends to multiply the effect of nitrous' additional oxygen. Alcohol's sensitivity is high, and its ability to resist knock diminishes rapidly with higher temperatures, but liquid nitrous injection tends to keep the alcohol mixture cool. Together, the chilling effects of nitrous expansion and alcohol's high heat of vaporization produce markedly improved engine VE.

Interestingly, some experiments show that alcohol motors make more power with nitrous and gasoline injection, while gasoline engines make more power with nitrous and alcohol injection. The thinking is that straight alcohol-nitrous mixtures may be too cold to burn most effectively in a naturally aspirated motor, and that colder temperatures hinder the dissociation of nitrous that makes power-boosting oxygen available for combustion. Given the lower stoichiometric ratios of ethanol and methanol, the fuel flow rate of a nitrous system designed to deliver gasoline would have to be increased 230 percent when injecting alcohol and nitrous.

Nitromethane and Nitrous

Nitromethane's overwhelming advantage as a fuel is its large oxygen content and, its high energy nitrogen bonding. Other than this, nitromethane is a terrible fuel, prone to detonation and, in general, very hard on an engine. Since nitrous exaggerates the weakness of a fuel, nitrous increases nitromethane's tendency to detonate. Since nitro is often used in a blend with methanol or gasoline, it would not be accurate to say nitrous and nitromethane are incompatible. But both nitrous and nitromethane have the potential to severely damage engines, and when brought together, the margin of error is small. Results from the few experts who have the knowledge and experience to effectively experiment with nitro fuel blends and nitrous-enhanced air seem to indicate that raising nitrous levels and lowering nitro percentages is safer than the other way around. In fact, nitropropane or nitroethane, which have less oxygen,

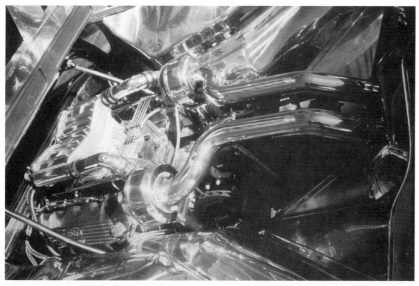
New owners of Megabuck later decided to convert the port injection system to nitrous/fuel injection. Such port injection systems can easily add several hundred horsepower to a big-block engine.

may be better candidates for nitrous injection than nitromethane. For many engines, the higher energy content per pound of a high methanol-nitrous blend will probably produce equal or better results with nitrous injection than a fuel blend very rich in nitromethane without nitrous injection.

Although nitrous had historically been banned due to some accidents, by the mid-1980s some sanctioning bodies were allowing nitrous-boosted alcohol cars to run in Top-Fuel drag classes. This is attractive to some racers, given that a Fueler requires in round figures $300 worth of fuel for each run while an alcohol car would only require $30 in nitrous to equal the heating value and oxygen content of the nitromethane.

David Vizard points out that although the 8-71 Roots-type blower engine with nitromethane fueling owns top-fuel racing, turbocharged Cosworth engines with only 132 cubic inches easily achieved 1,500 horsepower on alcohol and were designed to last 500 miles. Extrapolating this to the 500-plus cubes typical of Top-Fuel drag engines, in theory the potential exists to combine nitrous, alcohol, and turbocharging to build drag cars that launch well due to the nitrous boost, and then quickly build 3,000–6,000 horsepower for the realm beyond 100 yards where dragsters begin to become power-limited. The increased efficiency of large turbochargers combined with transient nitrous boost to provide a low-end kick and spool the turbos up fast has been used extremely effectively on street supercars like the Norwood Autocraft twin-turbo/nitrous Ferrari Testarosas.

Interestingly, while nitrous was banned from top-fuel drag racing, some Fueler engine builders were injecting small amounts of nitrous at idle to help prevent plug fouling and produce a cleaner idle on nitro.

Gaseous Fuels and Nitrous

Nitrous injection can be made to work with natural gas and propane. The biggest problem with gaseous fuels is the complete lack of liquid fuels' cooling effect as liquid fuels boil into a gas. Nitrous-propane or nitrous-natural gas tends to produce really hot combustion, the only ameliorating factor being the high resistance to knock of both fuels. Mechanical engine components and systems must be designed to withstand extreme heat, and water injection should be considered to help reduce combustion temperatures; otherwise nitrous injection with gaseous fuels will not be safely capable of generating power levels anywhere comparable to those possible with liquid fuels. By contrast, where liquid fuels are being used, it is superfluous to design in both excess cooling fuel and water injection.

There has been some recent experimentation of computer-controlled liquid propane injection, using the latest solenoid technology available to introduce liquid propane into an engine with a set-up which is essentially throttle body injection. The liquid propane injected will undergo rapid expansion and cooling.

Tests of supercharged WWII piston-engine aircraft with nitrous injection indicate that on detonation-limited engines, trading off small amounts of gaseous nitrous injection for supercharger boost resulted in net power gains on nearly a linear basis with percentage replacements of inlet air by nitrous oxide. Ten percent nitrous, with 2 psi less boost in an engine on the threshold of detonation improved power ten percent before detonation again became the limiting factor. The cooling effect of excess fuel was considered critical in this experimentation. David Vizard suggests that at the high levels of nitrous injection typical of automotive nitrous systems, the results of the above aviation testing with minor amounts of nitrous are probably not very relevant.

Dual-Fuel Injection

14

Inventors have long recognized the advantages of a vehicle that could run on multiple fuels. Pickup trucks in eastern Arizona 20 years ago frequently had both gasoline carbs and Impco Propane carbs—one mounted on top of the other! Gasoline started the vehicle more easily, made more power, and you could get it at any gas station. But propane was way cheaper. A dashboard switch manually activated solenoids that could effect fuel changeover at any time.

That was in the days of carbs. The advent of programmable electronic fuel injection has given the concept of dual-fuel vehicles new life. Dallas supercar builder Bob Norwood has applied the dual-fuel concept to a recent supercar project which involved turbocharging an Acura NSX. By retaining the stock internal engine parameters, including 10.2:1 compression ratio, the Norwood turbo NSX retains the stock fuel injection system under all driving conditions, yielding excellent drivability including great low-end crispness and torque, plus emissions compliance and the economy and convenience of street gasoline. Under turbo boost, an auxiliary Haltech F3 computer activates two huge 1600cc Indy car fuel injectors which inject super-high-octane methanol for air-fuel mixture enrichment and detonation control, enabling the car to safely run 9 psi boost. The NSX runs an auxiliary 5-gallon methanol tank and a totally separate fuel supply to a high-pressure fuel rail, which supports the two single-point injectors, each of which fires three times per engine revolution near the throttle body. The F3 computer controlling the additional injectors is

Dual-fuel Mazda Miata utilizes an entirely separate Haltech programmable and four auxiliary fuel injectors to provide enrichment fuel for an add-on turbo system. Under normally aspirated conditions, 87 octane fuel is sufficient, but for high turbo boost, racing gas enrichment via the auxilliary dual-fuel system will enable such a system to run higher levels of turbo boost without excessive ignition timing retard.

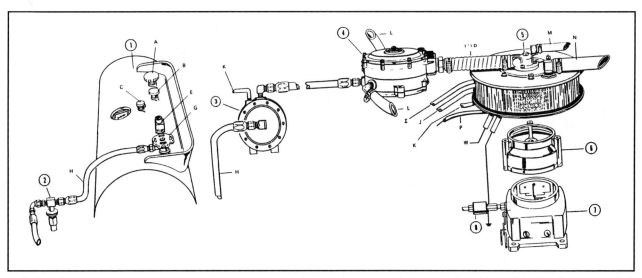

A dual fuel system mounts a propane mixer atop the original equipment carburetor. Solenoids enable the driver to switch back and forth between gasoline and propane fueling, taking advantage of propane economy and low emissions where available, while retaining the flexibility to utilize gasoline when necessary. *Impco Carburetion, Inc.*

virtually "sensorless," reading only engine speed and manifold pressure to provide timed injection. The car looks normal from the outside, idles normally, but step on the gas, she goes wild.

Norwood has now extended the dual-fuel injection concept to other vehicles with intermittent injection of high-octane 106 MON race gas, and even nitromethane. The dual-fuel injection scheme makes sense from many points of view: exotic fuel cost and availability, detonation control, emissions compliance, improved power, and improved dynamic range of injection.

Consider emissions. The automotive aftermarket has been feverishly working to build legal add-on or modified parts with legal California emissions Exemption Orders.

This is one reason auxiliary computers and injectors enter the picture: It is often far simpler to use one or more auxiliary computers to modify the behavior of the factory engine management system than to start from scratch building an entirely new set of engine control parameters with the precision required to pass emissions testing. Since many performance modifications only make a difference at full throttle (which is not part of the Federal Test Procedure for emissions compliance), the add-on auxiliary computer is a potent weapon in the hands of 1990s tuners. And when you have secondary injectors, it is relatively easy to supply them with a different fuel.

There are power advantages to dual-fuel injection. The fact that methanol contains 10 percent oxygen helps support additional combustion—a little like shooting nitrous oxide into a motor. Nitromethane contains even more oxygen, acting nearly like a monopropellant, which is why nitro can be incredibly dangerous. Multiple-injector Norwood auxiliary fueling systems plumb additional injectors into individual runners and provide an entirely separate fuel supply. Norwood has used up to three

Ferrari Testarosa from Norwood Autocraft has 12 additional injectors blowing high-octane gasoline into the intake ports downstream of the intercoolers.

"Charlie Brown III" rat motorboat engine utilizes entirely separate fueling for enrichment under boost conditions. Boat uses "river gas" for normal running, but the Haltech computer provides staged injection of high-octane racing gasoline for "lake burnouts."

electronic injectors per cylinder for nitro or alcohol, producing prodigious amounts of power on demand, while offering precise computer control of both fuels.

If emissions compliance is not necessary, electronic controllers like the F3 can stage dual-fuel injection, providing basic engine fuel control under all circumstances, while activating an additional set of dual fuel injectors at a specific rpm or manifold pressure. This has the added benefit of extending the dynamic fueling range of the stock injectors to much higher power levels while allowing the engine to idle properly at the other end of the scale. Norwood "tunes" the dual-fuel mix of such a system by balancing the size of the primary and secondary injector nozzles, since all injectors are

The Acura NSX supercar uses a high output V-6 motor with triple-lobe cams. When early NSX power was eclipsed by the competition, NSX owners began looking for ways to pump the NSX.

Norwood Autocraft installed a single turbo system on this NSX, retaining the stock injection for drivability and emissions legality. High output turbocharging of the NSX is a challenge due to the car's high compression ratio.

Norwood turbo NSX has a dual-fuel approach to supplying fuel for turbo boost conditions. Twin 1600cc/min Indy car-type methanol injectors under Haltech F-3 computer control come into play. Methanol injection provides an intercooling effect that radically cools incoming air, helping to keep combustion temperatures down. Methanol also provides the benefit that it is an extremely high-octane blending agent for gasoline. High-octane alcohols typically offer even higher octane as a blending component.

currently limited to injecting fuel for an equal length of time when controlled by a single computer. An auxiliary computer, as on the turbo NSX, has no such limitation, firing dual-fuel injectors entirely independently. Norwood has used dual-fuel injection on naturally aspirated engines, switching at a pre-programmed rpm entirely to alternative fuel, for example 3,000–4,000rpm in a circle track car. A Norwood twin-turbo dual-use powerboat normally runs on river gas, with a separate tank of 106 race gas for lake "burnouts."

Exotic fuel availability can be a serious problem for any dual-use vehicle, even if you don't care what it costs. Vehicles like powerboats and club-race cars do not require super-high-octane fuel except under heavy loading conditions. Running on pump gas vastly extends the practicality of such a car or boat. At the same time, controlling detonation is essential. Racers buy race gas and run it constantly, while street cars are happy on 87 or 92 octane. For the lunatic fringe that require a dual-use street-mannered vehicle with incredible horsepower on demand, dual-fuel injection is a free lunch.

15

Gear Oils, ATFs, and Other High-Performance Fluids

This chapter discusses certain high-performance fluids for automotive use, including synthetic and mineral oils and additives for lubricating manual transmissions, automatic transmission fluids, anti-freeze and coolant performance improvers, and other miscellaneous automotive fluids of interest to the enthusiast or automotive racer.

Gear Lubricants

Gear lubricants have essentially the same purpose as a motor oil, although the task is easier in some ways, and more difficult in others, than that of crankcase lubricants. Gear oils do not have to deal with harmful blow-by fuel contaminants as do crankcase lubricants (although they must be able to contain other contaminants). On the other hand, gear lubricants require enhanced friction-reducing and heat-removal abilities. Gear lubricants use the same type of oxidation, rust, and foam inhibitors as crankcase lubricants, although gear oil oxidation is different from that of crankcase oil, causing oil thickening that can eventually raise viscosity to the point that the oil becomes rubber-like. However, the antiwear and extreme pressure additive package must contain components that are active when needed, while remaining otherwise inert. Reconciling high speed with high torque can be a problem, because some materials that enhance high speed lubrication can degrade high torque performance. Extensive real-world and lab testing is required to sort out the effective and harmless materials. A recent trend has been to improve scoring protection without sacrificing thermal and oxidative stability by using sulfur-phosphorous lubricants. Rear-axle lubricants in some cases need to provide lubrication for limited-slip differentials, which have become increasingly sensitive recently to lubricant frictional properties. The current interest in the fuel economy of heavy commercial vehicles has led to testing of lower viscosity gear lubricants such as SAE 80W-140, which improved fuel economy 2.8 percent compared to SAE 85W-140. The U.S. military requires a single multi-purpose gear lubricant that meets both passenger car and truck requirements.

Manual transmission gear oils must facilitate easy shifting in cold weather. They must keep moving metal surfaces separated, reduce friction and wear, prevent scoring or welding of highly stressed parts, and prevent localized high temperatures. The viscosity must be such that at a specified low temperature the viscosity remains below 150,000 centipoise, the point at which gears can no longer be shifted. The gear lube must prevent "channeling," in which the gear oil is not replaced quickly enough as the film of lubrication on a tooth is wiped away by the companion tooth. Channeling gets its name from the fact that a gear tooth cuts a "channel" through the lubricant. Excessive wax in a gear lubricant can cause channeling even if the lubricant was of proper viscosity. Gear oils are designed with a very low pour point to prevent this.

The hypoid axle is the most difficult gear train to lubricate. The hypoid gear teeth are subjected to extreme pressure as high as 400,000 pounds per square inch while a high degree of sliding of the teeth over each other is occurring. Metal-to-metal contact is possible if the oil film ruptures. High relative gear

	SAE Viscosity Number							MIL-L-2105C Specification (a)		
	70W(e)	75W	80W	85W	90	140	250	75W(b)	80W-90	85W-140
Vis. @ 100°C (c) Min (cSt)	4.1	4.1	7.0	11.0	13.5	24.0	41.0	4.1	13.5	24.0
Max (cSt)	no req.	no req.	no req.	no req.	<24.0	<41.0	no req.	—	<24.0	<41.0
Max Temp. For Vis. of 150,000 cP,°C (d)	−55	−40	−26	−12	no req.	no req.	no req.	−40	−26	−12
Channel Point, Min °C	no req.	no req.	no req.	no req.	no req.	no req.	no req.	−45	−35	−20
Flash Point, Min °C	no req.	no req.	no req.	no req.	no req.	no req.	no req.	150	165	180

Values to be reported for: Gravity, API; Pour Point; Viscosity Index; Pentane Insolubles; Phosphorus; Chlorine; Nitrogen; Organo-metallic components; Sulfur: Finished Oil

Notes: (a) The MIL-L-2105C Specification replaced the MIL-L-2105B Specification
(b) The MIL-L-2105C 75W Classification replaces the MIL-L-10324A Sub-arctic Specification
(c) Viscosities determined by ASTM D 445 procedure
(d) Viscosities determined by ASTM D 2983 procedure
(e) Proposed

Physical requirements for gear lubricants. *Lubrizol*

Material	Density g/cm^3	Thermal Conductivity Watt/m·°C	Thermal Convection Watt/m^2·°C	Heat Capacity cal/g·°C	Heat of Vaporization cal/g
Water	1.000	0.60	1829	1.000	539
Glycol	1.114	0.25	-----	0.573	226
50/50	1.059	0.41	897	0.836	374
Aluminum	2.70	155		0.225	
Cast Iron	7.25	58		0.119	
Copper	8.93	384		0.093	
Brass	8.40	113		0.091	
Ceramics		1-10			
Air	.0013	.026		0.240	

Thermal properties of cooling system materials. Water is a better thermal transfer medium than ethylene glycol antifreeze, which reduces the cooling system efficiency in return for boil-over and freeze protection. *Red Line*

rubbing can cause excessive wear even where tooth pressures are insufficient to cause scoring and welding, which can occur in hypoid and worm gears. The extreme pressure (EP) additives form compounds on the gear teeth that will not weld, and are also called anti-weld or antiseize additives. Many gear oils are formulated for both hypoid-type applications as well as conventional transmission lubrication, although some manufacturers recommend against using these multipurpose lubricants with EP additives in their transmissions.

The concentration of gear oil additives is no greater than those in crankcase oil, varying from a few parts per million to several percent.

Gear lubricant additives include:
1) Oxidation inhibitors
2) Rust corrosion inhibitors
3) Oiliness or antiwear compounds
4) Friction-reducing additives
5) Extreme pressure agent
6) Antifoaming agents

Most of the standards for the above have been provided over the years by U.S. Military Specifications MIL-L-2105 and MIL-L-2105D.

Automatic Transmission Fluids

Automatic transmissions have become increasingly important not only in street automobiles, but in racing vehicles, as well as in farm and construction equipment, which have seen increasing use of manually selected power-shifting transmissions. Automatic transmission fluids (ATF) must:
1) Transmit power through the torque convertor
2) Function as fluid in the hydraulic control system
3) Transfer heat
4) Lubricate transmission clutches, gears, bearings, and seals
5) Provide friction modification ("oiliness")

ATFs include the following additives to base lubrication stocks to meet the above functionality:

Additives
1) Anti-oxidants
2) Dispersants
3) Antifoam agents
4) Extreme pressure additives
5) Friction modifiers
6) Viscosity temperature modifiers
7) Pour-point depressants
8) Seal conditioners
9) Corrosion inhibitors

A critical problem in ATF fluid performance is the ability to support necessary clutch pack frictional

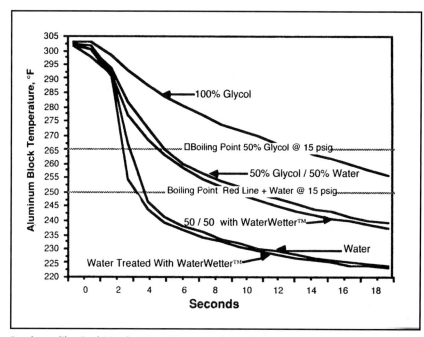

Products like Red Line's WaterWetter reduce the surface tension of water, enabling tiny steam bubbles to break loose from the cooling jacket surface, thus improving heat transfer. *SAE*

PROPERTY	RED LINE	SPEC	COOLANT A
pH	8.5	7.5 - 11	9.8
Boiling Point @ 15 psig	250°F		265°F (50%)
Freezing Point	31°F	-35°F(50%)	-35°F
Foaming Height, ml	5	150	50
Color	Pink	green/blue	green
Ash, %	0.5	5, max	1
Surface Tension @ 100°C, Dynes/cm^2	28.3	58.9 (water)	
Simulated Service Test, Weight Loss, mg/specimen			
Copper	4	20 max	11
Solder (low-lead)	0	60	1
Solder (high-lead)	9	---	---
Brass	4	20	11
Steel	1	20	6
Cast Iron	1	20	12
Aluminum	+1	60	0

WaterWetter contains corrosion inhibitor additives. Unlike antifreeze, there is no freeze protection, and boil-over protection is provided by high cooling system pressures. *SAE*

characteristics while meeting the other fluid performance properties. As always, this sort of trade-off is a different story on a street vehicle than on a race car where performance is everything and longevity is routinely traded for improved performance in the short run.

There are several different breeds of ATF for street vehicles due to differences in transmission clutch frictional properties. Where one design may require a slippery fluid with a low coefficient of friction at lockup in order to get a smooth shift without the problem of "stick-slip" noise and wear, another requires a higher coefficient of friction to provide fast clutch plate lockup that prevents wear from excessive slippage. General Motors specification ATF is called Dexron Automatic Transmission Fluid. Ford-approved ATF is either Mercon or Type F.

The ATF fluid must maintain its properties over extended periods. Any oxidation can cause sludge and varnish formation which could ultimately cause the transmission to fail from sticking of hydraulic controls and clogging of passages and orifices.

The ATF must also maintain sufficient viscosity at high temperatures to prevent hydraulic and control system leakage which would produce undesirable changes in shifting behavior due to excessive pump, pressure cylinder, and control valve leakage. ATF can reach temperatures of over 400 degrees F due to ATF's function as a power-transmitting medium in the torque converter. ATF must have a high viscosity in order to function at these temperatures. Excessive viscosity at low temperatures is equally undesirable, reducing ATF flow and so entailing efficiency loss in fluid members, extended shift time, and reduced low-temperature starting capability. As with crankcase oils, proper mineral oil or synthetic base stock and the proper additives for VI improvement and oxidation prevention get the job done. In addition, the fluid must be compatible with all elastomers and other transmission seals and components. The ATF must also provide high flash and fire points. Ordinary ATF is good to about -40 degrees F.

High-Performance ATF

Products like Racing Transmission Fluid and others are designed to change the hydraulic and clutch friction characteristics of ATF in order to produce more positive shifts, as well as to stand up to extremely high heat without oxidizing or aerating.

High-Performance Gear Lubricants

Unocal, Sunoco, Red Line, and others offer racing gear oil, which may or may not be synthetic. These are designed to stand up to severe shock loading and excessive heat generated during racing use. Sunoco claims its Racing 80W-90 Gear Oil has a unique borate chemistry that delivers a film thickness substantially greater than conventional gear oils, which allows lower viscosity and increased thermal transfer properties—temperature reductions of 50 degrees F have been reported. The oil is dyed blue to aid diagnosis

SAE Viscosity Grade	Maximum Temperature For Viscosity of 150,000 cP °C	Viscosity at 100°C	cSt
		Minimum	Maximum
70W	-55	4.1	-
75W	-40	4.1	-
80W	-26	7.0	-
85W	-12	11.0	-
90	-	13.5	less than 24.0
140	-	24.0	less than 41.0
250	-	41.0	-

Note: 1 cP = mPa•s; 1 cSt = 1 mm^2/s
Extracted from SAE Recommended Practice J306

Axle and manual transmission lubricant viscosity classification. Viscosity is directly related to ease of manual shifting, which can literally become impossible with high-vis lubricants in very cold temperatures.

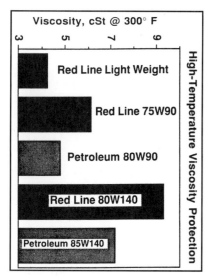

Synthetic gear oils like Red Line 75W-90 typically add about 5 percent axle efficiency compared to 85W-140 mineral oil lubricants. *SAE*

of gear contact patterns. Unocal's racing gear lube 90 is a multipurpose automotive gear lubricant compounded with lots of extreme pressure additive. It is designed to meet severe lubrication requirements in racing applications and is recommended for all high-offset hypoid gears in rear axles in racing vehicles.

High-Performance Coolants and Additives

Red Line Synthetic Oil Corporation's WaterWetter is designed to improve metal wetting along with corrosion protection as an additive to plain water or glycol coolants. Corrosion protection is especially critical for high-performance engines, which often include aluminum heads and/or blocks. Aluminum has a very high corrosion potential which is only abated by the fact that aluminum can form a protective film of oxidation, which prevents further corrosion. Poor aluminum corrosion protection allows dissolution of aluminum heat rejection surfaces, according to Red Line, weakening cooling system walls, water pump casings, and head gasket mating surfaces. The products of corrosion typically form deposits on lower temperature surfaces like radiator surfaces. The deposits have poor heat transfer capabilities, which dramatically reduce the efficiency of the cooling system.

Plain water has twice the heat transfer capability of a 50 percent water-glycol mixture, and most street vehicles have sufficient cooling capacity using such a solution. Products like WaterWetter in racing vehicles may allow a smaller radiator, with lower drag. The product reduces head temperatures, even compared to water alone, by doubling the wetting ability of water by breaking down surface tension. By improving coolant efficiency, head temperatures are reduced, reducing the possibility of detonation and allowing more spark advance for increased torque. Inlet system temperatures can also be reduced in some engines, for reduced detonation and improved volumetric efficiency. High-performance vehicles with marginal cooling may gain better hot-weather capability. In addition to surface-tension reducers, WaterWetter also contains additives to inhibit rust, corrosion, and metal electrolysis, to clean and lubricate water pump seals, prevent foaming, reduce cavitation corrosion, and reduce water scaling.

By reducing the surface tension of water, a product like WaterWetter allows water vapor to release more easily from metal surfaces. This can be important, since under heavy loading conditions much of the combustion heat in a cylinder head is transferred into the cooling system from boiling at localized hot spots, even though most of the coolant may be below the boiling point.

Brake Fluids

Being non-compressible, hydraulic brake fluids serve as a transmission medium to actuate the vehicle's wheel cylinders. Brake fluids can get very hot under extreme braking conditions; an important design criteria is resistance to boiling. On the other hand, brake fluids must function perfectly under the most extreme winter conditions, so the pour point must be extremely low temperature. Government standards for brake fluids are designed

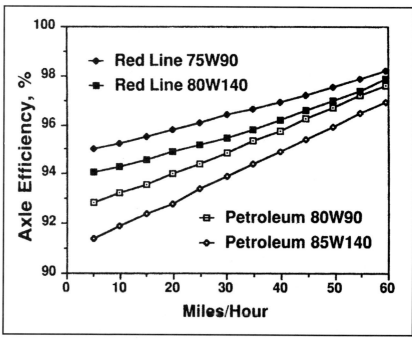

High-temperature viscosity protection comparison of Red Line gear lubes and petroleum 80W-90. *SAE*

to make sure all commercial brake fluids are safe, and street vehicles should always use DOT 3 or 4 brake fluid. There are two problems with ordinary brake fluid. The first is that brake fluid is hygroscopic, which means that it readily absorbs water from the atmosphere, which can eventually rust or corrode brake system components. Given that brake master cylinders are vented to the atmosphere, it is impossible to prevent some water absorption; brake fluid opened but not immediately added to the vehicle can absorb water in humid climates. The second problem is that brake fluid attacks paint; any spilled fluid can damage a vehicle's appearance. The solution is silicon-based brake fluid, which is obtainable at many ordinary auto parts stores. It does not damage paint, and it does not absorb water like ordinary fluid.

Brake fluids of glycol-ether fluids are the most common type. The boiling point of DOT 4 brake fluids drops more slowly than that of DOT 3, and the DOT 4 hydroxyl (OH) groups are partially esterified with boric acid so as to be chemically reactive with water, and therefore capable of neutralizing the effects of moisture. Mineral oil fluids have an advantage in that they are not hygroscopic, but they cannot be mixed with glycol-ether fluids, since the combination can destroy elastomer brake components. Silicone fluids are also not hygroscopic, but they are more compressible and do not lubricate as well as other fluids.

Greases

Greases consist of a hydrocarbon base thickened with a metallic "soap." Together these elements form a variety of special-purpose greases that provide oil-type thin- or thick-film lubrication. Unlike oils, greases are designed to stick to moving parts while handling extreme pressures, and they do not drain from the contact location. Greases consist of a base mineral oil (paraffinic, napthenic, or aromatic) or synthetic oil (olefin polymers, alkyl aromatics, esters, silicones, fluorinated hydrocarbons, or perflouropolymers) plus a thickener soap (Li, Na, Ca, Ba, or Al) organic thickener, or inorganic thickener, plus additives for extreme pressure, wear-protection, friction modifiers, adhesion improvers, oxidation inhibitors, and solid lubricants (molybdenum disulphide or graphite).

Where extreme heat is common, and a grease is required which must not contaminate other parts by melting and flowing (such as on disc brake components), synthetic greases are recommended. Synthetic greases have been used for the wheel bearings of race cars.

Other greases are available for very specialized applications. The lubrication of boat trailer wheel bearings, which are immersed in salt water and must inhibit rust and corrosion, is one application. As an example, Sta-Lube sells retail grease products including Boat Trailer W.B. (wheel bearing) Grease, Heavy Duty W.B. Grease, Moly-Graph Grease, E.P., M.P. (with molybdenum and graphite for resisting scuffing, scoring, and seizure in extreme pressure applications), Super White Grease for general purpose applications, Disc-Brake Hi-Temp W.B. Grease, C.V. Joint Grease, Sta-Plex Premium Red E.P. Grease (for industrial equipment with shock loads and extreme pressure), Lithium G.P. Grease (light duty, economical), Synthetic Brake Caliper Grease (with moly, graphite, and Teflon), and E.P. Anti-Seize Engine Assembly Lube (stays put over long storage periods and protects parts at initial start up).

Light Lubricants

There are many "household" light lubricants which consist of low viscosity hydrocarbons, with or without additives. WD-40, one of the most widely known, is largely kerosene (hydrocarbon similar in size to jet fuel and diesel oils) and molybdenum disulphide (MoS_2). WD-40 is great for light lubrication at room temperatures, and the moly additive remains even when the kerosene evaporates. Hydrocarbons between diesel and motor oil sizes (in the 10-25 carbon range) have the low viscosity required for applications like lubricating bike or cycle brake and throttle cables, while resisting sticking.

Cleaning Fluids and Solvents

Brake cleaner—one of the best solvents around for removing grease—consists of aromatic hydrocarbons such as toluene (C_7H_8), alcohols such as methanol (CH_3OH), and other "petroleum distillates." Brake cleaners sometimes contain chlorinated hydrocarbons (ethyl chloride, or C_2H_5Cl), which smell terrible and are even more vigorous than ordinary brake cleaner at dissolving grease and heavier hydrocarbons. Xylene and toluene—two high-octane aromatic components of many racing gasolines—are also commonly found in lacquer thinner and other paint solvents. Acetone $(CH_3)_2CO$, a dry-cleaning fluid, is another common solvent and is commonly used to clean up fiberglas resins. Methylethyl ketone, which is extremely hazardous to breathe, is another common solvent which has been used to clean up airplane cements and other adhesives.

Glossary

Abnormal combustion Combustion in which knock, pre-ignition, run-on, or surface ignition occurs, i.e., combustion which does not proceed in the normal way (where the flame front is initiated by the spark and proceeds throughout the combustion chamber smoothly and without detonation).

Additive Material added in small amounts to finished petroleum products to improve certain properties or characteristics.

Adsorption The adhesion by weak forces of materials to the surface of solid bodies in which they are in contact often as a monomolecular film, although the layer can sometimes be two or more molecules thick.

Advance When used in relation to the timing of a spark for initiating combustion, it is the number of degrees of crankshaft rotation that the spark fires earlier than a fixed or optimum setting. It is often expressed as degrees before top dead center (BTDC).

Air-fuel ratio The proportions, by weight, of air and fuel supplied for combustion.

Alcohols A group of colorless organic compounds, each of which contains a hydroxyl group. The simplest alcohol is methanol, CH_3OH.

Alkane A hydrocarbon having the general formula CnH_{2n+2}. Also called a paraffin.

Alkylation A refinery process for producing high-octane components consisting mainly of branch-chain paraffins. The process involves combining light olefins with isoparaffins in the presence of a strong acid catalyst such as sulfuric acid or hydrochloric acid.

Alkyl group A group of atoms, derived from an alkane (paraffin), having the general formula CnH_{2n+1}, which forms part of a molecule. Examples are the methyl group (CH_3), the ethyl group (C_2H_5), etc.

Alternative fuel An alternative to gasoline or diesel fuel which is not produced in a conventional way from crude oil.

Amide A compound containing the group $CONH_2$. The hydrogen atoms on the nitrogen can also be substituted by other groups.

Amine There are three types of amine. Primary amines are compounds containing the group NH_2 attached to an alkyl of aryl radical. Secondary amines have the group NH attached to two alkyl and/or aryl groups, and tertiary amines have three alkyl or aryl groups attached to the nitrogen atom.

Aniline Point The minimum temperature for complete mixing of equal volumes of aniline and the test sample when evaluated by ASTM D 611. Often used to provide an estimate of the aromatic hydrocarbon content of a mixture.

Antiknock additive An additive that, when added in small amounts to a gasoline, improves the octane quality of the fuel by suppressing knock.

Antiknock Index The average of the RON and MON for a fuel. Used as a measure of the octane quality of a gasoline, particularly in North America.

API gravity An arbitrary scale representing the gravity of density of liquid petroleum products in terms of API degrees, in accordance with the formula: API gravity (degrees) = (141.5/Sp Gravity 60/60 degrees F) - 131.5. The higher the API gravity, the lighter the material.

Aromatic A hydrocarbon based on a six-membered benzenoid ring.

Aryl SA hydrocarbon group containing a benzene ring where the benzene ring is directly attached to the rest of the molecule.

Auto-ignition The spontaneous ignition of a mixture of fuel and air without an ignition source. The auto-ignition temperature is the minimum temperature at which this takes place.

Aviation gasoline Special grades of gasoline produced for aircraft reciprocating engines. The antiknock quality is defined by a lean mixture rating (ASTM D 2700) and a supercharge rating (ASTM D 909). The vapor pressure is generally somewhat lower than for motor gasoline, and the distillation range can be narrower ideally from about 30 degrees C to about 150 degrees C.

Azeotrope A mixture of liquids whose distillation characteristics do not conform to Raoult's law, i.e., one that boils at a higher or lower temperature than any of its constituents. Azeotropic distillation is used as a method of separating materials having very similar boiling points, and which form suitable azeotropes.

Benzole A mixture of aromatic hydrocarbons containing a high proportion of benzene, obtained from the distillation of coal tar.

Biocides Additives used for killing microbiological growths, which often occur in the bottom of storage tanks, particularly those containing middle distillates.

Black products A generic term for any refinery stream containing residuum.

Black smoke Smoke and particulates emitted from a diesel engine when under load.

Blending number A value assigned to a compound or component that will enable it to be blended linearly with other fuel components so that the value of the finished blend can be predicted. They can refer to a number of properties such as octane, vapor pressure, etc.

Blow-by Fuel and gases that escape from the combustion chamber, past the pistons, into the crankcase.

Boiling range The spread of temperature over which a fuel, or other mixture of compounds, distills.

Bosch number A measure of diesel smoke determined by passing the exhaust gas through a white filter paper. The darkening of the paper is determined using a reflectometer, and Bosch numbers are reported on a scale from 0 (clear) to 10 (black).

Branch chain A description of a paraffinic hydrocarbon in which the carbon atoms are arranged in a branch form and not in a straight line.

Breakdown time Also known as the induction period. The time, in minutes, for a sample of gasoline when tested for oxidation stability by ASTM 525, to show a break point, i.e., when a pressure drop of 2 psi occurs over a period of 15 minutes in the bomb containing the sample and oxygen at 100 degrees C.

Brightstock Heavy lube oil obtained by refining residuum.

Carbonium ion A positively charged carbon radical in which the charge is due to the loss of one electron from the carbon atom.

Carburetor The device in some engine fuel systems that mixes fuel with air in the correct proportions and delivers this mixture to the intake manifold.

Carburetor foaming The formation of foam in a carburetor caused by the rapid boiling of fuel as it enters a hot carburetor. The foam cannot support the weight of the float so that more and more fuel enters the carburetor causing an increase in pressure and forcing excess fuel out through the vent and/or metering jets, so that an over-rich mixture is obtained.

Carburetor percolation This occurs when the fuel in a carburetor bowl starts to boil, either during or after a hot soak, forcing excess fuel into the inlet manifold via the vent or metering jet, so that an over-rich mixture results.

Catalyst A substance that influences the speed and direction of a chemical reaction without itself undergoing any significant change.

Catalytic converter A device in the exhaust system of an engine containing a catalyst in which reactions can occur that convert undesirable compounds in the exhaust gas into harmless gases.

Catalytic cracking A refinery process in which heavy hydrocarbon streams are broken down into lighter streams by the use of a catalyst and high temperatures.

Catalytic desulfurization A refinery process in which sulfur is removed from a hydrocarbon stream by combining it with hydrogen in the presence of a catalyst and then stripping out the hydrogen sulfide thus formed.

Catalytic reforming A refinery process which converts low-octane quality naptha to a high-octane blendstock (catalytic reformate) in the presence of a catalyst, mainly by converting napthenes and paraffins into aromatics. There are many commercially licensed versions of this process.

Caustic soda Sodium hydroxide (NaOH), a strongly alkaline chemical.

Cetane A paraffinic hydrocarbon, hexadecane ($C_{16}H_{34}$). The straight-chain isomer, n-cetane or n-hexadecane, is a primary reference fuel on which the cetane number scale for measuring the ignition quality of diesel fuels is based. It has a cetane number of 100. The other reference fuel is 2, 2, 4, 4, 6, 8, 8 heptamethyl nonane which has a cetane number of 15.

Cetane Index An approximation of cetane number based on an empirical relationship with API gravity and volatility parameters such as the mid-boiling point.

Cetane number A measure of the ignition quality of diesel fuel based on ignition delay in an engine. The higher the cetane number, the shorter the ignition delay, and the better the ignition quality.

Chassis dynamometer Equipment used to measure the power output of a vehicle at the drive wheels.

Climate chamber A room or chamber, usually containing a chassis dynamometer, in which various climatic conditions can be reproduced. Temperature control is most commonly used, but humidity, air pressure, sunshine, and rain can also be reproduced in a repeatable manner.

Cloud point The temperature at which a sample of a petroleum product just shows a cloud or haze of wax crystals when it is cooled under standard test conditions, as defined in ASTM D 2500.

CNG Compressed natural gas.

Coal tar One of the products of the destructive distillation of coal, the other main products being gas and coke.

Coking A refinery process that is an extreme form of thermal cracking, in which fuel oil is converted to lighter boiling liquids and coke.

Cold filter plugging point (CFPP) A measure of the ability of a diesel fuel to operate satisfactorily under cold weather conditions. The test measures the lowest temperature at which wax separating out of a sample can stop or seriously reduce the flow of fuel through a standard filter under standard test conditions.

Compression ignition The form of ignition that initiates combustion in a diesel engine. The rapid compression of air within the cylinders generates the heat required to ignite the fuel as it is injected.

Compression ratio The volume of the cylinder and combustion chamber when the piston is at BDC divided by the volume when the piston is at TDC.

Conjugated olefin An organic compound with alternate double and single bonds, e.g. 1, 3 butadiene $CH_2=CHCH=CH_2$.

Conversion process A process that converts heavy products to lighter products.

CFR (Cooperative Fuel Research) engine A single cylinder, overhead valve, variable compression ratio engine used for measuring octane or cetane quality.

Corrosion inhibitor An additive used in a fuel or other liquid that protects metal surfaces from corrosion.

Cracking A type of refinery process that involves converting large, heavy molecules into lighter, lower boiling ones.

Crude oil Naturally occurring hydrocarbon fluid containing small amounts of nitrogen, sulfur, oxygen, and other materials. Crude oils from different areas can vary enormously.

Cyclic dispersion The cycle-to-cycle variations in cylinder pressure that occur when an engine is running under otherwise constant conditions.

Dehydrogenation A chemical reaction that involves removing hydrogen atoms from alkanes or napthenes to give olefins or aromatics.

Demerit rating A numerical system in which increasingly high numbers represent increasingly poor performance. It is often used in evaluating vehicle drivability or the cleanliness of engine parts.

Demulsifier An additive used for breaking oil in water emulsions.

Density Mass of a substance per unit volume.

Delay period See ignition delay.

Detergent A fuel detergent is an oil-soluble surfactant additive that maintains the cleanliness of engine parts by solubilizing deposits or materials likely to deposit in the engine fuel system.

Detonation Often used to describe the uncontrolled explosion of the last portion of a fuel in the combustion chamber. See also knock.

Diesel index An obsolescent measure of ignition quality in a diesel engine, defined as: Diesel Index = 0.01(Aniline point)(API gravity)

Diesel knock An abnormal form of combustion that occurs in a compression-ignition engine and which is associated with long ignition delays. Often occurs when the engine is cold.

Direct injection (DI) A type of diesel engine in which the fuel is injected directly into the cylinder.

Dispersant A surfactant additive designed to hold particulate matter dispersed in a suspension.

Distillation The general process of vaporizing liquids in a closed vessel, condensing the vapors, and collecting the condensed liquids. Since the liquids vaporize generally in order of their boiling points, it provides a method of separating materials according to their volatility.

DON Distribution octane number a measure of the way octane quality is distributed across the boiling range of a gasoline, as measured using a modified CFR engine.

Drag coefficient A measure of the air resistance of a vehicle as it is being driven.

Drag reducing agent A fuel additive that can reduce resistance to flow so that the capacity of a pipeline is increased.

Drivability The response of a vehicle to the throttle. Good drivability requires such characteristics as smoothness of idle, ease of starting when hot or cold, smoothness during acceleration or hesitations or stumbles, and absence of surging at constant throttle when cruising. Separate tests are used for hot weather and cold weather drivability, in which numerical assessments of performance are assigned to each type of malfunction.

Elastomer Synthetic rubber-type materials frequently used in vehicle fuel systems.

Emulsification The formation of a dispersion of one liquid in another where the liquids are not miscible with each other, such as oil in water. Emulsifying agents, which are surfactant materials, will stabilize such emulsions and prevent them from separating into two layers.

Ethers A class of organic compounds containing an oxygen atom linked to two groups, which can be alkyl and/or aryl.

Evaporative loss controls Devices used on vehicles, service station pumps, tanks, etc., to prevent losses of light hydrocarbons by evaporation. Such controls on vehicles usually involve the use of a canister filled with activated charcoal into which vents from the fuel tanks, carburetor, etc., are fed so that the vapors are adsorbed on the charcoal. The canister is regenerated by drawing intake air through the canister while the engine is running so that the hydrocarbons are desorbed and burned in the engine.

Exhaust gas recirculation (EGR) The cycling of some exhaust gas back into the inlet manifold so as to lower combustion temperature and, hence, reduce nitrogen-oxide emissions.

FIA hydrocarbon analysis A fluorescent indicator adsorption test procedure defined by AST Procedure D1319 for the determination of hydrocarbon type in terms of aromatics, olefins, and saturates. The sample is adsorbed on silica gel containing a mixture of fluorescent dyes and is then desorbed down an activated silica gel column. The hydrocarbons are separated by types, and their positions indicated under ultraviolet light.

Flammability limits Mixtures of air and petroleum vapors will only burn or explode within a certain range of concentrations. The lean limit (or lower explosive limit) is where the mixture has just enough hydrocarbons to burn, and the rich limit (or upper explosive limit) is where it is almost too rich to burn.

Flash point The lowest temperature at which vapors from a petroleum product will ignite on application of a small flame under the standard test conditions.

Fluidizer A high-boiling-point, thermally stable organic liquid used as an additive in gasoline to reduce deposits on the undersides of intake valves.

Fractionation The separation by boiling point of mixtures of compounds by a distillation process. The degree of separation, i.e., the fractionation efficiency, will depend on the design of the fractionating tower.

Free radical A radical is a group of atoms, such as the methyl group (CH_3), that is part of a larger molecule. Such a group will not normally exist on its own since it has a free electron, but when it does it is called a free radical, and it will react with other materials such as oxygen, sometimes forming further free radicals.

Freezing point The temperature, determined under standard conditions, at which crystals of hydrocarbons formed on cooling disappear when the temperature of the fuel is allowed to rise.

Fuel injector A device for injecting fuel into a piston engine. They are used in all diesel and some gasoline engines where they replace the carburetor.

Fuel oil A term usually applied to a heavy residual fuel, although it can also be applied to heavy distillates.

Fungibility The ability to interchange or mix products from different sources without interactions occurring.

Gasohol A blend of 90 percent gasoline and 10 percent ethanol used as an automotive fuel in some states in the U.S.

Glow plug A plug-type electrical heater used as a cold starting aid in indirect injection diesel engines.

Gum The oxidation product arising from the storage of automotive fuel such as gasoline. It is barely soluble in gasoline and so will separate out and form a sludge. In an engine it will form sticky deposits which can cause malfunctions.

Hartridge unit A measurement of the black smoke emitted from a diesel engine in which the opacity of the smoke is determined.

Heat of combustion Also called the thermal, calorific, or heating value, and refers to the heat liberated when a fuel is burned. The upper or gross heating value includes the latent heat of water from combustion that is condensed in the test procedure. In an engine, the water is exhausted in the vapor form, and so a correction is made to give the net or lower heating value by subtracting the latent heat of condensation of any water produced.

Heat of vaporization Also called latent heat of vaporization. The heat associated with the change of phase from liquid to vapor at a constant temperature.

Heptane An alkane or paraffin having seven carbon atoms with the formula C_7H_{16}. Normal heptane, in which the carbons are arranged in a straight chain, is a primary reference fuel with the research and motor octane values of zero. There are nine isomers of heptane altogether.

Hydrocarbon A compound made up of hydrogen and carbon only.

Hydrocracking A refinery process in which heavy streams are cracked in the presence of hydrogen to yield high-quality distillates and gasoline streams.

Hydrodesulfurization A refinery process in which sulfur is removed from petroleum streams by treating it with hydrogen to form hydrogen sulfide, which can be removed from the oil as a gas by stripping.

Hydrofining A proprietary name for one version of the hydrodesulfurization process.

Hydrogenation Treatment of a stream with hydrogen, usually to remove sulfur or to stabilize it by saturating double bonds.

Hydrophilic group An organic group that has an affinity for water, such as the hydroxyl group (OH) or an acid group.

Hydrophobic group The opposite of hydrophilic. Most hydrocarbon groups are in this category.

Hydroskimming refinery A simple refinery consisting only of process units for distilling, catalytically reforming, and hydrotreating.

Ignition delay The period between the start of injection and the ignition of a fuel in a diesel engine.

Indirect injection (IDI) A type of diesel engine in which the fuel is sprayed into a prechamber to initiate combustion, rather than directly into the cylinder, as in a direct injection engine.

Induction period See breakdown time.

Injectors See fuel injectors.

Intake system icing The formation of ice in the carburetor or parts of some injector systems of spark-ignition engines that can cause vehicle malfunctions such as stalling and loss of power. It occurs only during cool humid weather when there is enough moisture in the intake air to condense and then freeze in the carburetor due to the temperature drop caused by the evaporation of the gasoline. If the ambient temperature is too low (below about 0 degrees C), there is not enough water in the atmosphere to give icing, and if the temperature is too high (above about 15 degrees C), then the temperature depression is not enough to freeze the water.

Isomers Compounds that have the same composition in terms of the elements present but which have the individual atoms arranged in different ways. Thus, there are two isomers of butane: n-butane (CH_3-CH_2-CH_2-CH_3) and isobutane (CH_3-CH-CH_3).

Isomerization A refinery pro-cess that converts normal or straight-chain hydrocarbons that have a poor octane quality into high-octane branch-chain isomers. Thus n-butane is converted into isobutane, etc.

Iso-octane The hydrocarbon 2, 2, 4-trimethylpentane that has eight carbon atoms and is used as a primary reference fuel with assigned values of RON and MON 100.

Kerosene (or Kerosine) A refined petroleum distillate of which different grades are used as a lamp oil, as heating oil, and as a fuel for aviation turbine engines.

Knock In a spark-ignition engine, auto-ignition (sometimes called detonation) of the end gas in the combustion chamber creates the characteristic knocking or pinging sound. It can cause damage to the engine and can be overcome by increasing the octane quality of the fuel or by engine modifications. In diesel engines, it is caused by excessive pressures in the combustion chamber and is avoided by the use of higher cetane number fuels.

Knock sensor A detector, usually fixed to the cylinder head, that detects when knock is occurring in a spark-ignition engine and actuates a mechanism such as one that retards the ignition to overcome the knock.

Latent heat The heat associated with a change of phase, such as going from a solid to a liquid at constant temperature (latent heat of fusion) or from a liquid to a gas at constant temperature (latent heat of vaporization or, simply, heat of vaporization).

Lead alkyl A class of head compounds, most commonly with methyl and/or ethyl groups attached to the lead atom, that are used as antiknock compounds in gasoline. See also TEL and TML.

Lead antagonism Some compounds, and particularly those containing sulfur, are antagonistic to lead in that, when they are present in a gasoline, they reduce the antiknock benefit given by lead alkyls. The sulfur compounds themselves vary, according to their chemical composition, in the degree to which they are antagonistic.

Lead response The extent to which the octane quality of a stream or component is improved by the addition of lead alkyls.

Lean mixture An air-fuel mixture that has an excess of air over the amount required to completely combust all the fuel.

Low temperature flow test (LTFT) A test to predict the low-temperature performance of a diesel fuel; gives a better correlation with field performance for U.S. vehicles and fuel than the cold filter plugging point (CFPP) test used in Europe and elsewhere.

LPG Liquefied petroleum gas that consists mainly of propane and/or butane, and which can be stored as a liquid under relatively low pressure for use as a fuel.

Markers Chemicals added to a fuel and which can be detected by a color reaction when in contact with another chemical. Used to detect theft, contamination, tax evasion, etc.

Mercaptans Compounds having the group -SH, also known as thiols, that have an extremely unpleasant odor and are removed from automotive fuel components to avoid customer complaints. Very low concentrations are often added to LPG to give it a distinctive warning odor.

Merit rating A numerical scale in which high numbers represent good performance. See also demerit rating.

Merox treating A proprietary refinery process for removing mercaptans from petroleum streams.

Metal deactivator A fuel additive that deactivates the catalytic oxidizing action of dissolved metals, notably copper, on hydrocarbons during storage.

Methanol Methyl alcohol, CH_3OH, the simplest of the alcohols. It has been used, together with some of the higher alcohols, as a high-octane gasoline component and is a useful automotive fuel in its own right.

Methyl tertiary butyl ether (MTBE) An oxygenated compound having the formula CH_3-O-C_4H_9, used widely as a high-octane component of gasoline.

Misfire Failure to ignite the air-fuel mixture in one or more cylinders without stalling the engine.

MMT Methylcyclopentadienyl manganese tricarbonyl, an antiknock additive sometimes used in conjunction with lead and sometimes used on its own in unleaded gasolines.

Motor octane number (MON) A measure of the antiknock quality of a fuel as measured by the ASTM D 2700 method. It is a guide to the antiknock performance of a fuel under relatively severe conditions as can occur under full throttle; i.e., when the inlet mixture temperature and the engine speed are both relatively high.

Multifunctional additive An additive or blend of additives having more than one function.

Naptha Loosely defined term covering a range of light petroleum distillates used as chemical and reformer feedstocks, gasoline blend components, solvents, etc.

Naphthenes A group of hydrocarbons having a cyclic structure with the general formula CnH_2n. Examples are cyclohexane and cyclopentane.

Natural gas A naturally occurring gas, consisting mainly of methane.

Naturally aspirated engine An engine in which the air entering the intake system is at atmospheric pressure.

Nozzle coking Deposit formation in the nozzle of a metering-type injector.

Octane number A measure of the antiknock performance of a gasoline or gasoline component; the higher the octane number, the greater the fuel's resistance to knock. There are two main types of octane number, the research octane number (RON), and the motor octane number (MON), which are based on different engine operating conditions and, therefore, relate to different types of driving mode. Both are based on the knocking tendencies of pure hydrocarbons; n-heptane has an assigned value of zero, and iso-octane a value of 100. The octane number of a fuel is the percentage of iso-octane in a blend with n-heptane that gives the same knock intensity as the fuel under test when evaluated under standard conditions in a standard engine.

Above a level of 100, the octane rating is based on the number of milliliters of tetraethyl lead per gallon, which is added to iso-octane to give the same knock intensity as the fuel under test.

Octane requirement The octane number of a reference fuel (which can be a primary reference fuel or a full boiling range fuel) that gives a trace knock level in an engine on a test bed or a vehicle on the road when being driven under specified conditions.

Octane requirement increase (ORI) The increase in octane requirement that occurs in an engine over the first several thousand miles of its life, due to the buildup of carbonaceous and other deposits in the combustion chamber. It is influenced by driving mode, gasoline composition, and the presence of lead.

Oil dilution The dilution of the lubricating oil by gasoline or partially combusted gasoline, which can find its way past the piston rings into the crankcase, particularly during cold starting. The lighter portions of the gasoline are evaporated off as the oil heats up, but the remaining material can reduce lubricating performance and, hence, increase wear.

Olefin An unsaturated hydrocarbon, that is, containing one or more double bonds.

Oleophilic group A chemical group attached to a molecule having an affinity for oily materials.

Oleophobic group The opposite of oleophilic.

Operability limit The lowest ambient temperature at which a diesel fuel will just function satisfactorily without wax separation, causing filter plugging problems.

Otto cycle The four-stroke cycle of most piston engines, i.e., intake, compression, power, and exhaust.

Oxidation Loosely, it is the chemical combination of oxygen with a molecule, although strictly it has a much broader meaning. It can be part of a manufacturing process or it can represent the deterioration of organic materials, such as gasoline or diesel fuel, due to the slow combination with oxygen from the air.

Oxygenate, oxygenated compound Terms that have come to mean compounds of hydrogen, carbon, and oxygen that can be added to gasoline to boost octane quality or to extend the volume of fuel available. Examples are methanol, ethanol, and methyl tertiary butyl ether (MTBE).

Paraffin A hydrocarbon having the general formula CnH_2n+2. Also called an alkane.

Particulates Particles, as opposed to gases, emitted from the exhaust systems of vehicles.

Pintle A type of fuel injector nozzle in which a shaped extension at the end of the injector needle controls the initial rate of fuel injection. When the needle is fully lifted, the fuel is full flow.

Pipestill The primary distillation equipment used in a refinery, in which the crude oil is heated in a furnace and passed into a fractioning tower, where it is split into different boiling range fractions.

Polymerization The combination of two or more molecules of the same type to form a single molecule having the same elements in the same proportions as in the original molecule, but with a higher molecular weight. The product is a polymer. The product of two or more dissimilar molecules is known as copolymerization.

Port fuel injection Fuel injectors that inject into the intake port rather than directly into the cylinders. They can be electronically or mechanically operated.

Pour point The lowest temperature at which a petroleum product will just flow when tested under standard conditions, as defined in ASTM D 97.

Preflame reaction The chemical reaction that takes place in an air-fuel mixture in an engine prior to ignition.

Pre-ignition The premature ignition of the fuel-air mixture in a combustion chamber, i.e., before the spark from the plug. It can be caused by glowing deposits, hot surfaces, or the auto-ignition of the fuel itself.

Primary reference fuel (PRF) For use in spark-ignition engines, it is a blend of n-heptane and iso-octane used as a primary standard for knock evaluations. The octane value (both RON and MON) or a primary reference fuel is the percentage of iso-octane in the blend with n-heptane.

For compression-ignition engines, primary reference fuels are used to define the cetane quality of a fuel and are usually n-cetane (cetane number of 100) and heptamethyl nonane (cetane number of 15).

Pyrolysis gasoline (pygas) Term frequently used to describe naptha produced by steam cracking and used as a gasoline component or gasoline.

Quench The removal of heat during combustion from the end gas or outside layers of the air-fuel mixture by the cooler walls of the combustion chamber.

R100 degrees C The research octane number of the part of a gasoline distilling up to 100 degrees C using a standard distillation apparatus (ASTM D 86). The difference between the RON of the whole fuel and the R100C is the delta R100 degrees C.

Reflux ratio A term used in connection with distillation and referring to the ratio of condensed sidestream returned to a distillation column to the amount taken off as a sidestream.

Reforming Sometimes used as a short form of catalytic reforming, which is a process for making high-octane components from naptha. Also, it is a mild thermal cracking process for naptha.

Refutas Chart A temperature-viscosity chart devised to show a linear relationship for Newtonian fluids.

Reid vapor pressure A measure of the vapor pressure of a liquid as measured by the ASTM D 323 procedure; usually applied to gasoline or gasoline components.

Repeatability The maximum difference between test results carried out on the same sample by the same operator using the same test equipment, above which the test is considered suspect.

Reproducibility The maximum difference between test results carried out on the same sample by different operators in different laboratories, above which the test is considered suspect.

Research octane number (RON) A measure of the antiknock quality of a gasoline as determined by the ASTM D 2699 method. It is a guide to the antiknock performance of a fuel when vehicles are operated under mild conditions such as at low speeds and low loads.

Residue, residuum The non-volatile portion of a crude oil resulting from distillation.

Response The way a fuel responds to treatment with additives such as lead alkyls, cold flow improvers, cetane improvers, etc., which are used to improve particular properties of the fuel.

Rich mixture A fuel-air mixture that has more fuel than the stoichiometric ratio.

Road octane number (RdON) Usually the octane number of a primary reference fuel (PRF) that just gives trace knock in a vehicle on the road or chassis dynamometer when tested under specified conditions.

Run-on A condition in which a spark-ignition engine continues to run after the ignition has been switched off. Also known as "after-running" or "dieseling."

Saturated compound A paraffinic hydrocarbon (alkane), i.e., a hydrocarbon with only single bonds and no double or triple bonds.

Scavanger Term applied to the halogen compounds (usually dibromoethane and/or dichloro-ethane) present in a lead antiknock compound to prevent lead compounds such as lead oxides and sulfates from building up in the combustion chamber.

Sensitivity The difference between the research octane number and the motor octane number of a gasoline. It is a measure of the sensitivity of the fuel to changes in the severity of operation of the engine.

Shale oil A largely hydrocarbon mixture derived from oil shale (a naturally occurring deposit) by distillation. It is a potential future alternative to crude oil.

SHED An acronym for "sealed housing for evaporative determination," a sealed chamber in which a vehicle is placed in order to determine the amount of evaporative losses that occur from its fuel system.

Silicone Organic compounds containing silicon, often used as antifoaming agents.

Solvent oil An alternative term for fluidizer.

Spark knock The most common form of knock, so called because it is influenced by the spark timing.

Squish The squeezing of part of the air-fuel mixture out of the end gas region in certain types of cylinder heads as the piston reaches the end of the compression stroke. It promotes turbulence and further mixing of the air and fuel and minimizes the tendency to knock.

Steam cracking A petrochemical process for the production of ethylene in which naptha is cracked in the presence of steam.

Stoichiometric air-fuel ratio The exact air-fuel ratio required to completely combust a fuel to water and carbon dioxide.

Stoke A unit of kinematic viscosity (the quotient of the dynamic viscosity and the density). One stoke is 1 cm2/s. A more convenient unit is the centistoke where 1 cSt = 0.01 St.

Storage stability The ability of a fuel to resist deterioration on storage due to oxidation.

Straight chain A descriptive term applied to a hydrocarbon in which all the carbon atoms are arranged consecutively in a straight line.

Supercharger A mechanical device that pressurizes the intake air or air-fuel mixture to an engine and so increases the amount delivered to the cylinders and, hence, the power output.

Surface ignition The ignition of the air-fuel mixture in a combustion chamber by a hot surface rather than by the spark.

Surfactant additive When applied to fuels, it is an organic compound with oleophobic and oleophilic groups that will form a coating on metal and other surfaces with the oleophilic group sticking into the hydrocarbon and the oleophobic group attaching itself to the surface. In this way it can protect surfaces and act as a detergent or dispersant by partially soulblizing deposits.

Susceptibility The extent to which the octane quality of a gasoline stream is improved by the addition of lead alkyls.

Sweetening process A refinery process for converting mercaptans (thiols) into nonodorous compounds.

Synthesis gas A mixture of carbon monoxide and hydrogen obtained from coke or natural gas by partial oxidation or steam reforming.

TEL See tetraethyl lead.

Terne plate Steel sheet coated with a lead/tin alloy and often used for the construction of vehicle fuel tanks.

Tertiary amyl methyl ether (TAME) An oxygenate used as a gasoline blend component.

Tetraethyl lead (TEL) A volatile lead compound, $Pb(C_2H_5)_4$, widely used as an antiknock additive in gasoline.

Thermal cracking A refinery process for converting heavy streams into lighter ones by heat treatment.

Thermostat A device used for the automatic regulation of temperature.

Toluene A relatively volatile aromatic compound, $CH_3C_6H_5$, present in catalytic reformate and widely used as a solvent. It has excellent octane qualities and can be used as a gasoline blend component.

Turbocharger A device, driven by exhaust gas pressure, for pressurizing the intake air or air-fuel charge of an engine, so as to increase the mixture delivered to the cylinders and, hence, increase power.

Unsaturated compounds Hydrocarbons having one or more double or triple bonds. Such compounds are reactive and will combine with other elements such as oxygen or hydrogen. Olefins are unsaturates.

Valve seat recession The wearing away of a valve seat in the cylinder head or an engine. Not usually a problem when lead is present in gasoline because the lead combustion products act as a lubricant. Engines designed for unleaded gasoline have hardened valve seats.

Vapor-liquid ratio The ratio, at a specified temperature and pressure, of the volume of vapor in equilibrium with the liquid to the volume of liquid charged, at a temperature of 0 degrees C. It can be measured using test procedure ASTM D 2533 and is used to define the tendency for a gasoline to vaporize in the fuel system of a vehicle.

Vapor lock A gasoline supply failure to the engine of a vehicle due to vaporization of the fuel preventing the pump from delivering an adequate supply of fuel. Factors favoring vapor lock are high ambient temperatures, volatile gasoline, and vehicle designs where heat from the engine can give high fuel-line temperatures.

Vapor lock index (VLI) An index that combines the Reid vapor pressure and the percentage evaporated at 70 degrees C of a gasoline, and is a measure of the likelihood of a gasoline to cause vapor lock in vehicles on the road.

Vapor pressure The pressure exerted by the vapors derived from a liquid at a given temperature and pressure. For gasoline, the Reid vapor pressure, as determined using the test procedure ASTM D323, is used to define the vapor pressure of a gasoline.

Venturi In a carburetor it is the narrowing of the air passageway. This increases the velocity of the air moving through it and induces a vacuum responsible for the discharge of fuel through the jet into the airstream.

Visbreaking A refinery process for thermally cracking residual fuel oil, originally to reduce its viscosity, but now to produce cracked streams.

Viscosity A measure of the resistance to flow of a liquid.

Volatility The property of a liquid that defines its evaporation characteristics. Highly volatile liquids boil at low temperatures and evaporate rapidly.

Wankel engine A rotary engine in which a three-lobe rotor containing combustion chambers turns eccentrically in an oval chamber.

Wax High molecular weight, generally straight-chain paraffins having limited solubility in diesel fuel.

Wax antisettling additive (WASA) An additive that reduces the tendency for wax crystals to settle out on storage of diesel fuel.

Wax antisettling flow improver (WAFI) An additive that improves the cold flow characteristics of a diesel fuel and also reduces the tendency for wax crystals to settle out during storage.

Wax appearance point A measure of the likelihood of wax coming out of solution as temperature is reduced. It is the temperature at which separated wax just becomes visible when tested under standard conditions, as defined by ASTM D 3117, and gives a similar result to cloud point.

White product Products that do not contain any residue from distillation.

White smoke The smoke emitted during a cold start from a diesel engine, consisting largely of unburned fuel and particulate matter.

Wide cut A fraction from a distillation column that has a wide boiling range. Wide-cut fuels are of interest for both diesel and spark-ignition engines, particularly for military use, when a normal fuel might not be available.

Xylene An aromatic hydrocarbon $(CH_3)_2C_6H_5$, present in catalytic reformate and widely used as a solvent.

Appendix A
Summary of Data on the Knock Ratings of Hydrocarbons

(A.P.I. Research Project 45)

Hydrocarbon	Research octane number (CRC–F$_1$) ml TEL/US gal			Blending octane number	Motor octane number (CRC–F$_2$) ml TEL/US gal			Blending octane number
	0.0	1.0	3.0		0.0	1.0	3.0	
Alkanes								
n-Pentane	6.71	74.9	88.7	62	61.9	77.1	83.6	67
2-Methylbutane	92.3	+0.37	+1.0	99	90.3	100.0+		104
2,2-Dimethylpropane	85.5	97.4	+0.1	100	80.2	93.0	99.9	90
n-Hexane	24.8	43.4	65.3	19	26.0	51.1	65.2	22
2-Methylpentane	73.4	84.6	93.1	83	73.5	87.3	91.1	79
3-Methylpentane	74.5	85.0	93.4	86	74.3	87.5	91.3	81
2,2-Dimethylbutane	91.8	+0.1	+0.6	89	93.4	+0.6	+2.10	97
2,3-Dimethylbutane	+0.3	+2.1			94.3	+0.4	+1.79	107
n-Heptane	0.0	10.0	43.5	0	0.0	25.4	46.9	0
2-Methylhexane	42.4		73.2	41	46.4		74.5	42
3-Methylhexane	52.0		74.7	56	55.0		81.0	57
2,2-Dimethylpentane	92.8		+0.4	89	95.6		+2.43	93
2,3-Dimethylpentane	91.1	98.6	+0.3	87	88.5	99.0	+0.29	90
2,4-Dimethylpentane	83.1	93.7	96.6	77	83.8	93.0	99.1	78
2,2,3-Trimethylbutane (triptane)	+1.8			113	–0.1		+3.07	113
3-Ethylpentane	65.0	75.2	85.0	64	69.3	81.2	88.0	73
n-Octane			24.8	–19		0.7	28.1	–15
2-Methylheptane	21.7	34.4	57.6	13	23.8	45.0	61.4	24
3-Methylheptane	26.8	37.5	59.6	30	35.0	53.5	68.0	30
4-Methylheptane	26.7	38.7	61.1	31	39.0	55.4	70.1	48
3-Ethylhexane	33.5	46.3	61.1	49	52.4	65.9	80.0	49
2,2-Dimethylhexane	72.5	85.4	93.3	67	77.4	90.0	95.2	76
2,3-Dimethylhexane	71.3	82.5	91.7	71	78.9	88.4	93.7	76
2,4-Dimethylhexane	65.2	77.6	87.3	65	69.9	83.8	89.0	70
2,5-Dimethylhexane	55.5	68.0	81.6		55.7	71.6	82.9	
3,3-Dimethylhexane	75.5	86.2	94.6	73	83.4	95.4	+0.0	81
3,4-Dimethylhexane	76.3	88.4	94.7	67	81.7	92.5	7.1	80
2,2,3-Trimethylpentane	+1.2	+3.7		105	99.9	+0.7	+2.0	112
2,2,4-Trimethylpentane (*iso*-octane)	100.0	+1.0	+3.0	100	100.0	+1.0	+3.0	100
2,3,3-Trimethylpentane	+0.61	+2.7		100	99.4	+0.6	+1.9	110
2,3,4-Trimethylpentane	+0.22	+1.2		97	95.9	+0.2	+0.7	102
n-Nonane				–17				–20
2,2,5-Trimethylhexane				91				88
2,3,5-Trimethylhexane				81				78
2,2-Dimethyl-3-ethylpentane	+1.8			108	99.5		+0.8	112
2,4-Dimethyl-3-ethylpentane	+0.5			88	96.6		+0.4	93
2,2,3,3-Tetramethylpentane	+3.6			123	95.0		99.4	113
n-Decane				–41				–38
2,7-Dimethyloctane				20				20
4,5-Dimethyloctane				48				61
2,5,5-Trimethylheptane				31				41
3,3,5-Trimethylheptane				77	88.7		+0.2	87
2,2,3,3-Tetramethylhexane				126	92.4		96.7	112
Alkenes								
1-Butene				144				126
2-Butene				153				130
3-Methylpropene				170				139
1-Pentene	90.9		98.6	119	77.1	81.3	82.9	109
cis-2-Pentene				154				137
trans-2-Pentene				150				134
2-Methyl-1-butene	+0.2		+0.3	146	81.9		84.2	133
3-Methyl-1-butene				129				125
2-Methyl-2-butene	97.3	99.0	99.2	176	84.7	85.5	85.8	141

Notes: Figures prefixed by a + sign in the above table represent ml TEL/US gal of *iso*-octane to give an equivalent rating.

1.0 ml/US gal = 1.2 ml/UK gal = 0.264 ml/l
3.0 ml/US gal = 3.6 ml/UK gal = 0.793 ml/l

	ml TEL/US gal			Blending octane number	ml TEL/US gal			Blending octane number
Hydrocarbon	0.0	1.0	3.0		0.0	1.0	3.0	
Alkenes (contd.)								
1-Hexene	76.4		91.7	97	63.4		76.3	94
trans-2-Hexene	92.7		98.4	134	80.8		83.2	129
trans-3-Hexene	94.0			137	80.1		82.3	120
2-Methyl-1-pentene	95.1		99.3	126	78.9		86.7	114
3-Methyl-1-pentene	96.0		+0.05	113	81.2		82.2	114
4-Methyl-1-pentene	95.7		+0.5	112	80.9		84.5	108
2-Methyl-2-pentene	97.8	99.3	99.5	159	83.0	84.7	85.0	148
cis-3-Methyl-2-pentene				125				113
trans-3-Methyl-2-pentene	97.2		100	130	81.0		84.0	118
cis-4-Methyl-2-pentene	99.3			130	84.3			128
trans-4-Methyl-2-pentene	99.3			130	84.3			128
2-Ethyl-1-butene	98.3		+0.3	143	79.4		82.0	129
2,3-Dimethyl-1-butene				148				129
3,3-Dimethyl-1-butene	+1.7			137	93.5			121
2,3-Dimethyl-2-butene	97.4	98.1	98.5	185	80.5		84.0	144
1-Heptene				68				46
5-Methyl-1-hexene				96				80
2-Methyl-2-hexene	90.4			129	78.9			119
2,4-Dimethyl-1-pentene	99.2		+0.36	142	84.6		87.3	124
4,4-Dimethyl-1-pentene	+0.4		+0.8	144	85.4		87.7	136
2,3-Dimethyl-2-pentene	97.5		99.5	165	80.0		83.3	145
2,4-Dimethyl-2-pentene	100			135	86.0			123
2,3,3-Trimethyl-1-butene	+0.5	+0.7	+1.2	145	90.5	92.3	93.7	130
1-Octene	28.7	43.8	63.5		34.7	46.6	57.7	
2-Octene	56.3	69.9	78.7	75	56.5	67.9	73.0	68
trans-3-Octene	72.5	84.6	89.4	95	68.1	77.7	81.2	85
trans-4-Octene	73.3	85.4	91.8	99	74.3	82.8	84.2	101
2-Methyl-1-heptene	70.2	79.6	87.9	77	66.3	73.3	79.6	
6-Methyl-1-heptene	63.8	74.8	87.2	74	62.6	69.9	76.6	69
2-Methyl-2-heptene	75.9			91	71.0			102
6-Methyl-2-heptene	71.3	84.6	90.2	75	65.5	77.0	80.5	
2,3-Dimethyl-1-hexene	96.3			118	83.6	86.7	88.1	109
2,3-Dimethyl-2-hexene				144				
Di-*iso*-butylene	+0.5	+0.9	+1.1	168	88.6	89.1	90.1	151
2,3,3-Trimethyl-1-pentene	+0.6		+0.9	138	85.7		87.2	129
2,4,4-Trimethyl-1-pentene	+0.6		+1.0	164	86.5		88.8	153
2,3,4-Trimethyl-2-pentene	96.9		+0.02	142	80.9		84.4	130
2,4,4-Trimethyl-2-pentene	+0.3	+0.4	+0.6	148	86.2	87.3	88.0	139
3,4,4-Trimethyl-2-pentene	+0.3		+0.7	151	86.4		87.7	144
1-Nonene				35				22
Aromatics								
Benzene				99	+2.7			91
Methylbenzene (toluene)	+5.8			124	+0.3	+1.00	+1.7	112
Ethylbenzene	+0.8	+0.8	+0.8	124	97.9	100.0	+0.2	107
1,2-Dimethylbenzene (*o*-xylene)				120	100.0		+0.0	103
1,3-Dimethylbenzene (*m*-xylene)				145				124
1,4-Dimethylbenzene (*p*-xylene)				146				127
n-Propylbenzene	+1.5	+3.8	+4.3	127	98.7	+0.1	+0.2	129
iso-Propylbenzene (cumene)	+2.1	+3.4		132	99.3	+0.2	+0.5	124
1-Methyl-2-ethylbenzene				125				111
1-Methyl-3-ethylbenzene	+1.8			162	100			138
1-Methyl-4-ethylbenzene				155	97.0			115
1,2,3-Trimethylbenzene				118				105
1,2,4-Trimethylbenzene (*pseudo*-cumene)				148	+0.9			124
1,3,5-Trimethylbenzene (mesitylene)	>+6.0			171	+6.0			137

Notes: Figures prefixed by a + sign in the above table represent ml TEL/US gal of *iso*-octane to give an equivalent rating.

1.0 ml/US gal = 1.2 ml/UK gal = 0.264 ml/l
3.0 ml/US gal = 3.6 ml/UK gal = 0.793 ml/l

	Research octane number (CRC-F$_1$)				Motor octane number (CRC-F$_2$)			
	ml TEL/US gal			Blending octane number	ml TEL/US gal			Blending octane number
Hydrocarbon	0.0	1.0	3.0		0.0	1.0	3.0	
Aromatics (contd.)								
n-Butylbenzene				114	95.3			117
iso-Butylbenzene				122				118
sec-Butylbenzene				116				117
tert-Butylbenzene	>+3.0			138	+0.8			127
1-Methyl-4-n-propylbenzene				152				139
1-Methyl-3-iso-propylbenzene				154				136
1-Methyl-4-iso-propylbenzene (cymene)	+1.4			150	97.7			133
1,3-Diethylbenzene	>+3.0		>+3.0	155	97.0		+0.2	144
1,4-Diethylbenzene				151	96.4			138
1,2,3,5-Tetramethylbenzene				154	+0.2			128
Cycloalkanes								
1,1-Dimethyl*cyclo*propane				116				108
1,1,2-Trimethyl*cyclo*propane	+1.5		>+3.0	133	87.8		93.0	123
1,1-Diethyl*cyclo*propane				114				108
Ethyl*cyclo*butane	41.1			30	63.9			62
*Cyclo*pentane				141	85.0	91.4	95.2	141
Methyl*cyclo*pentane	91.3	99.5	+0.5	107	80.0	89.4	93.0	99
Ethyl*cyclo*pentane	67.2	72.3	79.5	75	61.2	72.7	80.7	67
cis-1,3-Dimethyl*cyclo*pentane	79.2		91.2	98	73.1		86.8	85
trans-1,3-Dimethyl*cyclo*pentane	80.6		93.2	91	72.6		87.1	75
Dimethyl*cyclo*pentane	84.2		95.9	96	76.9		88.7	86
n-Propyl*cyclo*pentane	31.2	43.1	59.8	27	28.1	43.3	60.5	27
iso-Propyl*cyclo*pentane	81.1	89.6	94.3	83	76.2	85.7	89.4	81
n-Butyl*cyclo*pentane	−3		29.6	−4	−2		36.7	5
iso-Butyl*cyclo*pentane	33.4	47.9	59.2	34	28.2	40.6	58.1	37
tert-Butyl*cyclo*pentane				112				112
*Cyclo*hexane	83.0	92.9	97.4	110	77.2	85.4	87.3	97
Methyl*cyclo*hexane	74.8	83.5	88.2	104	71.1	82.0	86.2	84
Ethyl*cyclo*hexane	46.5	54.0	65.1	43	40.8	52.3	65.4	39
cis-1,2-Dimethyl*cyclo*hexane	80.9	89.2	94.3	85	78.6	87.2	90.7	82
trans-1,2-Dimethyl*cyclo*hexane	80.9	89.8	94.5	85	78.7	87.3	90.8	84
cis-1,3-Dimethyl*cyclo*hexane	71.7				71.0			
trans-1,3-Dimethyl*cyclo*hexane	66.9	75.7	83.5	67	64.2	78.3	83.8	65
cis-1,4-Dimethyl*cyclo*hexane	67.2	78.0	84.7	68	68.2	80.0	85.0	66
trans-1,4-Dimethyl*cyclo*hexane	68.3	75.1	82.8	64	62.2	77.4	83.4	59
n-Propyl*cyclo*hexane	17.8	25.6	42.8		14.0	29.4	47.7	
iso-Propyl*cyclo*hexane	62.8	70.1	79.6	62	61.1	74.3	81.4	62
1,1,3-Trimethyl*cyclo*hexane	81.3	89.5	94.8	85	82.6	91.2	95.8	92
n-Butyl*cyclo*hexane			22.5	−8		4.4	25.3	−4
iso-Butyl*cyclo*hexane	33.7	42.4	56.4	39	28.9	40.3	58.3	39
sec-Butyl*cyclo*hexane	51.0	59.5	71.2	44	55.2	64.2	74.6	58
tert-Butyl*cyclo*hexane	98.5	+0.2	+0.8	116	89.2	92.3	96.3	101

Notes: Figures prefixed by a + sign in the above table represent ml TEL/US gal of *iso*-octane to give an equivalent rating.

1.0 ml/US gal = 1.2 ml/UK gal = 0.264 ml/l
3.0 ml/US gal = 3.6 ml/UK gal = 0.793 ml/l

Sources: API Research Project 45
Technical Data on Fuel by J.W. Rose and J.R. Cooper, 7th Edition, The British National Committee, World Energy Conference, London, 1977.

Appendix B

Racing Fuels Specifications

VP Hydrocarbons C-10, Typical Values

(Unleaded racing gasoline, safe for O_2 sensors and catalytic converters)

API Gravity	?
Specific Gravity	
.760 at 60 degrees F	
Distillation	
Initial BP	?
10%	?
50%	?
90%	?
Endpoint	?
Color	clear
Corrosion 3 hrs. @ 122 F	1A
Gum, mg/100 ml	<1
Lead GMS/gal	0
Phosphorus, Theories	0
Sulfur Wt. %	<.001
Temperatures @ VL=20 F	?
Reid Vapor Pressure, lbs	?
BTUs/lb	?
Motor Octane Number	96
Research Octane Number	104
(R+M)/2	100
Price/gal	?
Price/drum	?

VP Hydrocarbons Performance Unleaded, Typical Values

(Unleaded street-legal high-octane gasoline, safe for O_2 sensors and catalytic converters)

API Gravity	?
Specific Gravity	
.760 at 60 degrees F	
Distillation	
Initial BP	?
10%	?
50%	?
90%	?
Endpoint	?
Color	clear
Corrosion 3 hrs. @ 122 F	1A
Gum, mg/100 ml	<1
Lead GMS/gal	0
Phosphorus, Theories	0
Sulfur Wt. %	<.001
Temperatures @ VL=20 F	?
Reid Vapor Pressure, lbs	?
BTUs/lb	?
Motor Octane Number	?
Research Octane Number	?
(R+M)/2	100
Price/gal	?
Price/drum	?

VP Hydrocarbons VP Red, Typical Values

(Leaded fuel for race engines under 12.5:1 compression)

API Gravity	?
Specific Gravity	
.742 at 60 degrees F	
Distillation	
Initial BP	?
10%	?
50%	?
90%	?
Endpoint	?
Color	clear
Corrosion 3 hrs @ 122 F	1A
Gum, mg/100 ml	<1
Lead GMS/gal	0
Phosphorus, Theories	0
Sulfur Wt. %	<.001
Temperatures @ VL=20 F	?
Reid Vapor Pressure, lbs	?
BTUs/lb	?
Motor Octane Number	105
Research Octane Number	?
(R+M)/2	?
Price/gal	3.75
Price/drum	200

VP Hydrocarbons C-12, Typical Values

(For compression ratios to 14:1)

API Gravity	68.7
Specific Gravity	
.7086 at 60 degrees F	
Distillation	
Initial BP	98
10%	129
50%	196
90%	218
Endpoint	240
Color	Green
Corosion 3 hrs @ 122 F	1A
Gum, mg/100 ml	<1
Lead GMS/gal	4.2
Phosphorus, Theories	0
Sulfur Wt. %	<.001
Temperatures @ VL=20 F	136
Reid Vapor Pressures, lbs	7.75
BTUs/lb	18.834
Motor Octane Number	108
Research Octane Number	110
(R+M)/2	109
Price/gal	4.75
Price/drum	245.00

VP Hydrocarbons C-14, Typical Values

API Gravity	72.4
Specific Gravity	
.6940 at 60 degrees F	
Distillation	
Initial BP	130
10%	170
50%	200
90%	212
Endpoint	232
Color	yellow
Corrosion 3 hrs @ 122 F	1A
Gum, mg/100 ml	<1
Lead GMS/gal	4.2
Phosphorus, Theories	0
Oxidation stability, minutes	1,440+
Sulfur Wt. %	<.001
Temperatures @ VL=20 F	171
Reid Vapor Pressure, lbs	5.5
BTUs/lb	19,684
Motor Octane Number	114
Research Octane Number	116
(R+M)/2	115
Price/gal	5.75
Price/drum	310

VP Hydrocarbons C-15, Typical Values

(for ultra high-compression engines where VP C15 offers insufficient protection)

API Gravity	61.7
Specific Gravity	?
Distillation	
Initial BP	?
10%	?
50%	?
90%	?
Endpoint	?
Color	green
Corrosion 3hrs @ 122 F	1A
Gum, mg/100 ml	<1
Lead GMS/gal	4.2
Phosphorus, Theories	0
Oxidation stability, minutes	1,440+
Sulfur Wt. %	<.001
Temperatures @ VL=20 F	?
Reid Vapor Pressure, lbs	?
BTUs/lb	?
Motor Octane Number	117
Research Octane number	?
(R+M)/2	?
Price/gal	?
Price/drum	?

VP Hydrocarbons C-16, Typical Values

(For ultra high cylinder pressure applications, turbos, blowers, air racing, and nitrous)

API Gravity	61.7
Specific Gravity	
.7324 at 60 degrees F	
Distillation	
Initial BP	196
10%	202
50%	206
90%	212
Endpoint	220
Color	blue
Corrosion 3 hrs. @ 122 F	1A
Gum, mg/100 ml	<1
Lead GMS/gal	4.2
Phosphorus, Theories	0
Oxidation stability, minutes	1,440+
Sulfur Wt. %	<.001
Temperatures @ VL=20 F	233
Reid Vapor Pressure, lbs	1.5
BTUs/lb	18,765
Motor Octane Number	117
Research Octane Number	118
(R+M)/2	117.5
Price/gal	6.00
Price/drum	320

VP Hydrocarbons C-18, Typical Values

(For high-compression engines)

API Gravity	71.8
Specific Gravity	
0.696 at 60 degrees F	
Distillation	
Initial BP	109
10%	153
50%	208
90%	212
Endpoint	230
Color	yellow
Corrosion 3 hrs.@ 122 F	1A
Gum, mg/100 ml	0.4
Lead GMS/gal	4.23
Phosphorus, Theories	0
Oxidation stability, minutes	1,440+
Sulfur Wt. %	.017
Temperatures @ VL=20 F	156
Reid Vapor Pressure, lbs	?
BTUs/lb	?
Motor Octane Number	116
Research Octane Number	?
(R+M)/2	?
Price/gal	7.00
Price/drum	410

VP Hydrocarbons VP Marine, Typical Values

(Leaded fuel for high power and throttle response)

API Gravity	?
Specific Gravity	
.735 at 60 degrees F	
Distillation	
Initial BP	?
10%	?
50%	?
90%	?
Endpoint	?
Color	Red
Corrosion 3 hrs. @ 122 F	1A
Gum, mg/100 ml	<1
Lead GMS/gal	0
Phosphorus, Theories	0
Sulfur Wt. %	<.001
Temperatures @ VL=20 F	?
Reid Vapor Pressure, lbs	?
BTUs/lb	?
Motor Octane Number	96
Research Octane Number	?
(R+M)/2	?
Price/gal	?
Price/drum	?

VP Hydrocarbons VP Air Race, Typical Values

(Highest octane VP fuel, leaded fuel good with nitrous, turbos, blowers when C16 has insufficient antiknock protection)

API Gravity	?
Specific Gravity	
.710 at 60 degrees F	
Aromatic hydrocarbons	10 %
Manganese g/gal	.2
Distillation	
Initial BP	?
10%	?
50%	?
90%	?
Endpoint	?
Color	Blue
Corrosion 3 hrs. @ 122 F	1A
Gum, mg/100 ml	<1
Lead GMS/gal	0
Phosphorus, Theories	0
Sulfur Wt. %	<.001
Temperatures @ VL=20 F	?
Reid Vapor Pressure, lbs	?
BTUs/lb	?
Lean Knock F-3 Method	120.3
Rich Knock F-4 Method	157.2
Motor Octane Number	?
(R+M)/2	?
Price/gal	?
Price/drum	?

Sunoco GT OPRG Unleaded, Typical Values

API Gravity	54.7
Specific Gravity	
.760 at 60 degrees F	
Aromatic Content	35%
Distillation	
Initial BP	90
10%	150
50%	210
90%	220
Endpoint	230
Color	Clear
Copper strip corrosion	No. 1
Gum, mg/100 ml	1
Lead GMS/gal	0
Phosphorus, Theories	?
Sulfur Wt. %	<?
Temperatures @ VL=20 F	?

Reid Vapor Pressure, lbs	8.0
BTUs/lb	?
Motor Octane Number	95
Research Octane Number	105
(R+M)/2	100
Price/gal	?
Price/drum	?

Sunoco GT Standard Leaded, Typical Values

API Gravity	63.7
Specific Gravity	
0.725 at 60 degrees F	
Aromatic Content	15%
Distillation	
Initial BP	90
10%	160
50%	220
90%	245
Endpoint	360
Color	Purple
Copper strip corrosion	No. 1
Gum, mg/100 ml	1
Lead GMS/gal	?
Phosphorus, Theories	?
Sulfur Wt. %	<?
Temperatures @ VL=20 F	?
Reid Vapor Pressure, lbs	8.0
BTUs/lb	?
Motor Octane Number	105
Research Octane Number	115
(R+M)/2	110
Price/gal	?
Price/drum	?

Sunoco M85 Blended, Typical Values

API Gravity	47.6
Specific Gravity	
0.790 at 60 degrees F	
Aromatic Content	35%
Distillation	
Initial BP	98
10%	140
50%	145
90%	146
Endpoint	147
Color	Clear
Copper strip corrosion	No. 1
Gum, mg/100 ml	1
Lead GMS/gal	?
Phosphorus, Theories	?
Sulfur Wt. %	?
Temperatures @ VL=20 F	?
Reid Vapor Pressure, lbs	7.0
BTUs/lb	?
Motor Octane Number	N/A
Research Octane Number	N/A
(R+M)/2	N/A
Price/gal	?
Price/drum	?

Unocal 76 Unleaded Racing Gasoline, Typical Values

(For showroom stock and other events when unleaded is legal; distributed in California only in winter)

API Gravity	48
Specific Gravity	
0.788 at 60 degrees F	
Aromatic Content	?
Distillation	
Initial BP	96.8
10%	145
50%	213
90%	244
Endpoint	280
Color	Blue
Copper strip corrosion	1a
Gum, mg/100 ml	1.0
Lead GMS/gal	?
Phosphorus, Theories	?
Sulfur Wt. %	0
Temperature @ VL=43 F	63
Reid Vapor Pressure, lbs	6.8
BTUs/lb	?
Oxygen (wt. %)	2.2
Motor Octane Number	94
Research Octane Number	106
(R+M)/2	100
Road Octane Number	101
Price/gal	?
Price/drum	?

Unocal 76 Racing Gasoline-Leaded, Typical Values

API Gravity	63
Specific Gravity	
0.728 at 60 degrees F	
Aromatic Content	?
Distillation	
Initial BP	96.8
10%	145
50%	213
90%	244
Endpoint	363
Color	Red
Copper strip corrosion	1a
Gum, mg/100 ml	1.0
Lead GMS/gal	?
Phosphorus, Theories	?
Sulfur Wt. %	0
Temperature @ VL=43 F	63
Reid Vapor Pressure, lbs	6.5
BTUs/lb	?
Oxygen (wt. %)	2.2
Motor Octane Number	104
Research Octane Number	112
(R+M)/2	108
Road Octane Number	110
Aviation performance number	112/160
Price/gal	?
Price/drum	?

Index

Air-fuel ratio, 4, 5, 12, 14, 25-27, 29, 65, 67, 75, 81, 120
Alcohol, 4, 13, 14, 19, 22, 23, 32, 33, 36, 41-43, 48, 50, 59, 60, 65-69, 75, 93, 95, 103, 104, 107, 114, 118
ATF, 110, 111
Avgas, 7, 35, 36, 38, 45, 48-54, 57, 62

Brake cleaner, 113
Brake fluid, 113

Chemical reaction, 9
Cooling effect, 57
Crude oil, 15, 16, 38, 93, 119, 120

Diesel, 4, 8, 13, 19, 24, 35, 48, 56, 87-92, 115-119, 121
Distillation curve, 57, 58

EPA, 8
Ethanol, 4, 8, 13, 14, 43, 44, 66-71, 73, 93, 102, 119

Formula One, 55, 61
Fuel injection, 29

Gasoline, 4, 5, 7, 8, 10, 12-16, 19, 22-24, 27, 29, 31-45, 48, 51, 53-62, 64-72, 75, 79-82, 84, 85, 87-89, 93, 95, 98, 101-103, 105, 114, 116-121

Jet fuel, 50

Kerosene, 33

Methanol, 4, 7, 8, 13, 14, 29, 43, 59, 65-76, 93, 102, 103, 105, 113, 119
Mineral oil, 15, 19, 20, 21, 113
Multi-port injection, 29

NASCAR, 23, 29, 55
Natural gas, 4, 7, 8, 13, 15, 19, 24, 29, 32, 65, 66, 77, 79-82, 84, 85, 88, 104, 115, 119, 121
Nitroglycerin, 73

Nitrous oxide, 4, 99, 100, 106

Octane, 6-8, 12, 27-30, 32-36, 38, 39, 41, 42, 44, 45, 49-52, 54-56, 58, 60-63, 77, 79, 82, 90, 93, 96, 115, 116, 118, 119, 121
Oil additives, 20

Port fuel injection, 15, 44, 71, 119
Propane, 4, 7, 8, 10, 15, 19, 24, 29, 32, 67, 77, 79-82, 84, 85, 91, 104, 105, 118

Specific gravity, 57
Swirl, 25, 26, 31

Viscosity, 15, 17-20, 22-24, 90, 109-111, 113, 121
Volatility, 42, 62, 90, 121